The Lost Language of Plants

Medicine

As dreams are the healing songs
from the wilderness
of our unconscious—
So wild animals, wild plants, wild landscapes
are the healing dreams
from the deep singing mind
of the earth.

—Dale Pendell, *Living with Barbarians*

The Lost Language of Plants

The Ecological Importance of Plant Medicines to Life on Earth

Stephen Harrod Buhner

Chelsea Green Publishing
White River Junction, Vermont

Thanks is gratefully extended to the following publishers and authors for permission to reprint:

from *The Collected Poetry of Robinson Jeffers*, volume 1, 1920–1928. Tim Hunt,ed., copyright 1938, 1966 Donna Jeffers and Garth Jeffers. Used with permission.
from *Cultures of Habitat* by Gary Paul Nabhan, copyright 1997 Gary Paul Nabhan. Reprinted by permission of Counterpoint Press, a member of Perseus Groups, L.L.C.
from *The Dragon Who Never Sleeps: Verses for Zen Buddhist Practice* by Robert Aitken, copyright 1992 Robert Aitken. Reprinted with permission of Parallex Press, Berkeley, California.
from *House of Light* by Mary Oliver, copyright 1990 Mary Oliver. Reprinted by permission of Beacon Press, Boston.
from *The Kabir Book* by Robert Bly, copyright 1971, 1977 Robert Bly, copyright 1977 The Seventies Press. Reprinted by permission of Beacon Press, Boston.
from *Living With Barbarians: A Few Plant Poems* by Dale Pendell, copyright 1999 Dale Pendell. Published by Wild Ginger Press, Sebastopol, California. Used by permission of the author.
from *Mind and Nature* by Gregory Bateson, copyright 1979 Gregory Bateson. Used by permission of Dutton, a division of Penguin Putnam, Inc.
from *Pharmako/Poeia: Plant Powers, Poisons, and Herbcraft* by Dale Pendell, copyright 1995 Dale Pendell. Published by Mercury House, San Francisco, and reprinted by permission.
from *The Soul's Code* by James Hillman, copyright 1996 James Hillman. Used by permission of Random House, Inc.

Technical editor: David Hoffmann.
Designed by Ann Aspell.
Printed in Canada
First printing, February, 2002.

05 04 03 02 1 2 3 4 5

Library of Congress Cataloging-in-Publication Data
Buhner, Stephen Harrod.
The lost language of plants : the ecological importance of plant medicine to life on earth / Stephen Harrod Buhner.
p. cm.
Includes bibliographical references (p.)
ISBN 1-890132-88-8
1. Materia medica, Vegetable. 2. Medicinal plants. 3. Pharmaceutical industry—Environmental aspects. 4. Human ecology. I. Title.

RS164 .B785 2002
615'.32—dc21 2001053868

CHELSEA GREEN PUBLISHING COMPANY
Post Office Box 428
White River Junction, VT 05001
(800) 639-4099
www.chelseagreen.com

For my son, Benjamin

Contents

A Note to the Reader

The book you are now holding in your hands is designed to be a book of feelings as well as thoughts. Like a novel it is designed to take you on a journey, in this instance one that ends well—though as with most novels bad things do happen along the way. Like most nonfiction it is designed to explore a number of ideas, perhaps new ways of looking at things. But unlike the majority of nonfiction it is also designed to explore a range of feelings. Some of these feelings are evoked in the stories I share, others in the way the informational sections are arranged and presented. The feelings that emerge as you read the book are important. I do not believe we can solve the environmental problems facing us unless we develop our capacity for feeling and our empathy for other life-forms to the same degree that we have developed our facility of thought. Unfortunately, because thinking is given precedence over feeling in Western cultures, most environmental books neglect feelings altogether. Those that do explore feelings either simply evoke the despair and pain that comes with the recognition of environmental devastation, or the awe that can occur in powerful and healthy natural settings—as if only the two choices were available to us. Though the pain, despair, and awe are important, things (and people) are a great deal more complex than that.

And so, this book explores the complex, multidimensional, intricately interconnected, synergistic, living organism that we call Earth, and it is designed to be complex and multidimensional as well. It is designed to loop back on itself—like living organisms do; to contain multiple layers of communication—like living organisms do; and to contain something of the wildness of the world that it explores. As well, it examines some of the ways of thinking we have been developing over the past few centuries and the environmental devastation that is implicit in some of those ways of thinking—specifically the devastation inherent in technological medicine and pharmaceuticals. This book also explores other ways of thinking: ways that offer a more sustainable alternative than the mind-set we now use, ways of healing that are much older than technological medicine, ways that are deeply involved in the uses of plants as medicines, ways that are concerned with the aesthetic unity that underlies the ecosystems of Earth. For the millions of plant chemistries that are released into ecosystems are done so in response to environmental needs—to communications directed to plants—while pharmaceuticals are released into ecosystems in the billions of tons without any reason whatsoever. Both types of substances affect the same metabolic pathways in living organisms and so cause significant changes in the functioning of ecosystems. However, plant chemistries are filled with meaning; they are a language, pharmaceuticals are not. And so finally, this book delves into the meaning embedded within plant chemistry, the language of plants—a language human beings in the Western world lost knowledge of when we began to think so insistently with the analytical portions of our brains and quit thinking with other, more holistic parts of ourselves. Thus this book harkens back to the original teaching of plants and wild systems and to the first teachers we ever knew.

Fifteen years ago I had an odd dream. In it, a medicinal plant that I was interested in, an *Usnea* lichen that is ubiquitous on trees throughout the world, told me that while it was good for healing human lungs it was primarily a medicine for the lungs of the planet, the trees. When I awoke, I was amazed. It had never occurred to me in quite that way that plants have some life and purpose outside their use to human beings. But, of course, when I thought it over I realized that the human species has only been around for about a million years. A lichen that has been intimately connected to tree species for at least 150 million years must have been doing something all that time besides pining away for our emergence.

Acknowledgments

In no particular order or hierarchy of importance:

Andre Norton, Christian Daughton, Thomas Ternes, David Hoffmann, Dale Pendell, Mark A. Geyer, Lawrence Gilbert, Mr. S. Song (a.k.a. 7Song), James Duke, Rosemary Gladstar, Sparrow, John Seed, Susun Weed, Kat Harrison, Gary Paul Nabhan, Laurie Weiss, Robert Bly, Henry David Thoreau, Gregory Bateson, Brooke Medicine Eagle, Barry Lopez, David Ehrenfeld, David Orr, Buckminster Fuller, Richard Milton, Louis Pascal, Brian Martin, Nan DeGrove, Jon Weiss, James Lovelock, Richard Nelson, Lynn Margulis, Reed Noss, Toni Knapp, Steven Foster, Greg Smith, James Huckabay, Joseph Kiefer, and Trishuwa.

The Taste of Wild Water | 1

The stirring of their twigs against the dark, travelling sky
Is the oracle of earth

—Ted Hughes, "A Wind Flashes the Grass"

A flower,
A willow,
A fisherman
On a rock.

A ray of sun
On the river,
A bird
On the wing.

Halfway
Up the mountain
A priest slowly climbs to a shrine.

In the forest
A yellow leaf
Flutters and falls.

—Ho P'ei-you, Ch'ing Dynasty

I was eight the first time I tasted wild water.

I lay with my great-grandfather along the bank of a pond deep in rural Indiana. I was close enough that his smell came strongly to me, subtly shading and comforting the wild smells surrounding me. It is a smell that I have known since I have known smell, deeply encoded in my body-memory, stored away with other body-memories that are bound together with the origins of my life: the feel of his starched shirts against my newborn skin, the smell of his soap and cigarettes, and his voice a deep rumble into which is interwoven the communications of ancient generations and nineteenth-century life.

My great-grandmother is there in body-memory, too. Her voice throaty. She was overweight and chain-smoking, hard-drinking and peering nearsighted, heavy-bosomed, strongly perfumed. And always with her memory are smells of gravy and fried chicken, green beans boiled to nonexistence, the taste of too-sweet iced tea in great, sweating glass jugs. And the feeling of being loved deep into the soul till it matters not the shape of a person's body or the irregularities of their personality.

I remember journeys to their formal stately home in Columbus: dark and dangerous basements, attics filled with memories and old smells, mirrors that distorted new generations peering half-frightened into them, and overstuffed horsehair chairs that could not be sat upon without sliding off. And out back of the house my great-grandfather's office where patients from the nineteenth century came to see their doctor.

Those patients had little trust in the new generation of physicians that peered condescendingly down at them, talking as if to children, certain of their educational pedigree, all in clean-white-shining-characterless faces. They came instead to be with a fellow traveler through time, who had been filled with stories of Civil War, begun life on horses and in buggies, had seen them die in one world war and then another. One who used the old medicines and still looked at the tongue, who palpated the organs, and who said, "Say 'ahhh.'"

I sometimes went in the back door of that office—drawn always against my will. It stood darkly in the back of the garden, pulling on me, until I could resist no longer. Then uncertain, trembling, I would walk past the goldfish pond filled with ancient flashing carp, along the overgrown walkway, to the blond brick building with the dark brown door. Even touching the handle brought its smell to the nostrils: old metal and years of human touch. The knob rattled slightly, but it took a pressure to turn it, for the door fit tightly. All of my strength was necessary and my hand would just start to slip when the door would let go with a sudden "thunk" and pop open a little. More smells would come, smells held at bay behind blond brick, peeling brown door, metal latch and knob. Strange smells: of herbal medicines and chemicals and leather, of oak and old paper, of old sickness and pain, and of the years of human lives, which has a smell all its own.

The door creaked when I pulled it open, hinges protesting, and it scraped along the concrete walk. And when it was finally open there was always that terrible moment when I had to decide whether or not to step inside. In the dim light lay old boxes and metal things, worn-out brooms, and scrapes on the walls, and that slightly damp concrete with its own wetly penetrating smell. As I breathed it in I was caught up and carried away to some kind of life that I did not understand. It seemed so old and somehow so sad that I could hardly take breath. But none of these things came to me in words. They expressed themselves in the instinctive motions of flight, half turning to the door again to go, to get away, to feel once again the sunlight play over my body and fill up my lungs. And sometimes I gave it rein and left not-running, remembering to close the door, to push on it until it snicked into place. And somehow in the back of my mind I could hear a sad thing crying out to me—a thing that I could not hold, could not understand, could not bear.

But other times I would not turn and go, not give that silent fear expression or control over my body. I would reach out and turn on the light, and sometimes I would even close the door behind me and place myself firmly in that world. Then, trembling, I would step out into it and it would close about me firmly, insistently, irrevocably.

Along the right side of the room were counters and above them shelves. In the middle of the counter was a large sink, stained by fearful liquids over unimaginable years. On the shelves above were brown bottles leaning, filled, leaking, mysterious, and strange. The smell was strong: of herbal medicines, and chemicals, and water, and plants, and age. Along the left side of the room were shelves filled with boxes and bottles and metal devices and things that my memory provides no shapes for.

Once I had made it this far, it was seldom that I did not go on. There was a door ahead of me and through it the room where my great-grandfather saw his patients. It was a room that felt safe, touched by human life. There was a high leather table with shiny metal things at one end that looked like stirrups or the spurs used by cowboys. There was a desk with a chair and papers and the walls were clean and it smelled like hospitals and fear and hope all at the same time.

At the far wall there was another door, always closed. It carried on its surface a shiny patina of age, the knob turned silently and easily, and on the other side was the waiting room. There was a couch, its leather dark with years, its oak body rubbed black and smooth. There were oak chairs and end tables with magazines and table lamps and along the front length of the couch there was a low table and on it a large glass jar, perhaps a foot across. It was filled with shark teeth and I knew I could take off the top and reach inside and take some if I wanted. Sometimes I did and wondered about a fish that could have these kind of teeth, and sometimes I would think about falling in the ocean—but not often.

Eventually I would turn and retrace my steps and always that back room would challenge me. I would snake through it quickly, hearing soft footfalls behind me that I could not turn and face, and it always seemed that I got the door closed just in time. I would shake then and not understand why I had ventured in and it might be a month or a year before I did it again.

It was in Columbus that I was introduced to the mysteries of my great-grandfather's profession but in the Ohio forests at his country home that I was introduced to the mysteries of the human and of Earth and of the interior world that all human beings possess.

The house there was a hand-hewn oak barn built a century before. My great-grandfather had bought it, numbered the pieces, disassembled it, moved it to their country farm, and rebuilt it. I remember visits there, lying next to him in bed and hearing his stories. It did not matter what he said but only the sound and rhythms of his voice, his arm about me, and his smell sinking deep within me. I remember one night lying with him when the story was done and looking up from slitted eyelids to his face and knowing that the only thing I wanted in life was to be a man like this man. Only years later did I realize the Universe hears such requests and shapes irrevocable destiny from child thought. And even more years passed before I realized that there is a price to granted wishes.

Sometimes he would take me out into the woods on that farm and we would walk. There is a special kind of shadow that happens in deep woods that are old and have been left undisturbed. Underneath the canopy of ancient hardwood trees the greens are deeper, the soil blacker, the smells richer. And there is a shadow that is over everything, calling out that there is a deeper world than the human of which we are a part. Something came out of that place and entered my body. I felt more whole, more human, more loved, more a part of the world. And in some indefinable way I *knew* who I was.

I remember the particular way my great-grandfather walked through those woods; few walk that way now. I see it mostly in old pictures, in the stance of ancient ancestors, of Civil War veterans, of people long gone. That way of walking has a particular smell, a particular gait, a particular rhythm, a particular integration with Earth and plant and water. As we walked through those woods he would push aside a plant in his path. But it was not brusquely done. Rather he moved them from his path as if they were relatives he was setting aside. The soil was black with a bit of clay and it could be easily formed with the hands. A shovel would go in deeply and there were few stones. The roots of the plants entwined in that deep dark soil and our feet sank down a little as we walked—as if we were moving on the living tension of the soil—like the water spiders that skated on the pond where we fished. My great-grandfather's feet knew the tension of that soil—they expected it and the soles of his feet spoke to it, conversed with Earth, each step of the way.

When we reached the pond, we would lie for hours on its banks, the silence a blanket over us. Sometimes we would drop a word into the silence like a stone into the water and the word's meaning would send ripples through us until they ebbed, slowed, stopped. Still, even then I knew those words that I loved were unnecessary. For in our time together we were doing something without words that humans have done for millennia. As we lay with the smells and the sounds and the feels of that place deep inside something would leave his body and enter mine. I would breathe it into me as slowly as I breathed in his smell; something in my soul found purchase in it. It was a food without which I could not have become human. It is always passed in silence between the man and the boy, between the woman and the girl. It is handed down from one interior world to the next. Its essence penetrates the muscles of the body, the oxygen of the blood, the substance of the spirit. And this was the time in which I first tasted wild water.

A man's hand possesses touch and touching, softness and hardness, those deep veins on its back that can capture the eyes and will not let them go, and all his life written in the lines of his palm. I remember lying back upon my elbows with a piece of grass between my teeth and my great-grandfather leaning forward and cupping his hand, sliding it under the surface of that glassy pond. He lifted the shining surface to my lips and said, "Here, have you ever tasted this water?" I looked at him askance and caught the gleam in his eye, then bent my head and pursed my lips. I remember the translucency of that water, the tiny particles of dirt floating in its depths, and below it all his life written in the palm of his hand. The water was sweet and cooling; my body liked it. As I lifted my head, I caught the glint in his eye and he made that peculiar gritty sound with his teeth as he smiled that I loved so much. "Good isn't it?" he asked. And I remember nodding. And then we lay back down and that thing continued to come out of his body into mine.

Later my mother caught me drinking wild water and told me it would kill me and began to instill in me a fear of the wildness of nature. And later still, my great-grandfather died and my days began to be filled with TV dinners and the flickering, half-intimacy of television. The years passed and the voices of my ancestors began to fade from memory; I became used to the taste of domesticated water.[1]

It was long and long again before I tasted wild water once more, before the seeds that my great-grandfather and the land had laid within me began their slow growth. Even more years before I was no longer afraid wild water would kill me. The journey back to wild water is a long one—for our species, for each of us. In making that journey we must find a way to heal within ourselves a wound laid down long ago, a wound that came from a certain decision our species made, from a certain way of thinking—a wound that can be most easily distinguished when remembering puppies.

It is easy to remember what a puppy looks like, perhaps even to imagine one on the other side of the room you are in now. He is smelling the floor, looking around, filled with the newness of life as puppies often are.

He begins to walk across the floor. He doesn't see you yet—all of his attention is consumed with what he is smelling, what he is seeing. He is walking with that funny gait that puppies often have, their hind end slightly askew as if their rear legs are walking faster than the front. The puppy gets nearer and you give a little whistle and say, "Here boy, come here," and you whistle again.

The puppy looks up and sees you ("that's a good boy," you say) and his whole body begins to wag. "It's you," the puppy seems to be saying, "it's you!" And in that moment something passes between you and the puppy. It is as if something leaves your body and enters the puppy; as if something leaves the puppy and enters you. And the most important thing then is to touch the puppy, to pet him, to hold him. And the puppy seems to want nothing more than these things as well—perhaps, in addition, to lick your hands or your face.

This is an experience that nearly all people know, yet we have no word for it in our language. (*Love* is too broad in its possible meanings, too overworked.) Once, people experienced this exchange with everything on Earth. The experience was understood, expected, a natural part of human life—this deep interaction with the nonhuman world—this exchange of soul essence.

But it is not common now. Once shared with all life-forms such an exchange now happens only in our immediate families (if we are lucky), with our pets—dogs and cats or other domestic animals we may meet from time to time—and, if we have a green thumb, with the gardens, lawns, or plants that we grow. In that sense, it is also a perverted experience. Domesticated animals and plants and landscapes are only a shadow of what plants and animals and landscapes can be and are in the wild. What we share with them is only a shadow of what we can share, and historically have shared, with the wild. The emotive energy that comes from a wolf is of an entirely different magnitude than that from a dog.

The loss of this kind of engagement, with life in general and with wild life in particular, by its nature generates deep wounds, external and internal. No longer feeling or exchanging an emotional affinity with all life, we are no longer care-full with Earth, its landscapes, and its many life-forms. From our care-lessness our environment becomes impoverished. And our interior world is impoverished as well. That joyful feeling that comes into our bodies when we exchange soul essence and deep emotional affinity with a puppy once happened many times a day—with many species of life. It was a regular part of the historical life experience of our species on Earth—a primary type of exchange. We

---

Along the upper Amazon . . . there are yachaqs *who learn shamanism directly from the* doctores, *the plant teachers. These* yachaqs *are called* vegetalistas. . . . *[And] every true* vegetalista *has to meet* Sacha Runa, *the Forest Person, face to face. In the Jungle.*

Sacha Runa is like us, except that he lives in the forest.
Sacha Runa is the one who takes care of the animals—
the one who lets them out in the morning.

By jungle I really mean primary forest—forest that hasn't been cleared—that's where Sacha Runa lives. And it's through Sacha Runa that one is introduced to the plant teachers, the doctoritos. *Sometimes Sacha Runa will come out from behind a ceibo tree and say hello to a person who wasn't even looking for him.*

—Dale Pendell, PHARMAKO/POEIA

are now so far from such emotional engagement as a daily part of life that we no longer have a common word to describe its occurrence.

These interior and exterior wounds have come from the development of a certain way of seeing the world and our place in it (and the loss of the older ways it replaced). The development of what Gregory Bateson calls an epistemology: a way of knowing or seeing reality.

The Two Wounds | 2

. . . far are we from the forests of our rest
Where the wolf nature from maternal breast
Fed us with strong brown milk
Yet still our souls keep memories of that time
In Sylvan wildernesses, our soul's prime
Of wisdom, forests that were gods' abode.

—Edith Sitwell, *Elegy on Dead Fashion*

The tree which moves some to tears of joy
is in the Eyes of others
only a Green thing that stands in the way.

—William Blake, *The Letters*

One of the penalties of an ecological education
is that one lives alone in a world of wounds.

—Aldo Leopold, *A Sand County Almanac*

THE EXTERIOR AND INTERIOR WOUNDS THAT COME FROM NO LONGER sharing soul essence with the world around us are pervasive, though the interior one is more difficult to see. The exterior wound, however, is easily perceived; people have been discussing its appearance for hundreds of years. By now it has widened, deepened, become so severe that most people routinely acknowledge its existence. This wound is the logging of the rain forest, the pollution and destruction of rivers . . . all the desecration of our exterior world. It has been talked about so much, and we have become so inured, that it is easy to forget that there is a *feeling* to this exterior wound. A feeling before words, before thinking. A simple, deep response from somewhere inside us recognizing damage to the fabric of life. We can shut these feelings off. But to understand the impact of the exterior and interior wounds it is important to feel them—even if only briefly—even if it hurts.

Henry David Thoreau described the exterior wound in his journal on December 30, 1851.

> This afternoon, being on Fair Haven Hill, I heard the sound of a saw, and soon after from the Cliff saw two men sawing down a noble pine beneath, about forty rods off. I resolved to watch it till it fell, the last of a dozen or more which were left when the forest was cut and for fifteen years have waved in solitary majesty over the sprout-land. I saw them like beavers or insects gnawing at the trunk of this noble tree, the diminutive manikins with their cross-cut saw which could scarcely span it. It towered up a hundred feet as I afterward found by measurement, one of the tallest probably in the township and straight as an arrow, but slanting a little toward the hillside, its top seen against the frozen river and the hills of Conantum. I watch closely to see when it begins to move. Now the sawers stop, and with an axe open a little on the side toward which it leans, that it may break the faster. And now their saw goes again. Now surely it is going; it is inclined one quarter of the quadrant, and, breathless, I expect its crashing fall. But no, I was mistaken; it has not moved an inch; it stands at the same angle as at first. It is fifteen minutes yet to its fall. Still its

> branches wave in the wind, as if it were destined to stand for a century, and the wind soughs through its needles as of yore; it is still a forest tree, the most majestic tree that waves over Musketaquid. The silvery sheen of the sunlight is reflected from its needles; it still affords an inaccessible crotch for the squirrel's nest; not a lichen has forsaken its mast-like stem, its raking mast,—the hill is the hulk. Now, now's the moment! The manikins at its base are fleeing from their crime. They have dropped the guilty saw and axe. How slowly and majestically it starts! as if it were only swayed by a summer breeze, and would return without a sigh to its location in the air. And now it fans the hillside with its fall, and it lies down to its bed in the valley, from which it is never to rise, as softly as a feather, folding its green mantle about it like a warrior, as if, tired of standing, it embraced the earth with silent joy, returning its elements to the dust again. But hark! there you only saw, but did not hear. There now comes up a deafening crash to these rocks, advertising you that even trees do not die without a groan. It rushes to embrace the earth, and mingle its elements with the dust. And now all is still once more and forever, both to eye and ear.

There is a feeling to this. A certain grief that is felt in the body, a clutch that overtakes the heart.

Things of course have not gotten better but worse in the intervening 150 years. Though we learn to dampen it as best we can, not a day goes by that we do not in some way feel this wound to our world, even in the smallest of things.

The poet Dale Pendell, in his book *Living With Barbarians,* published this description of it in the last year of the twentieth century:

> I had got used to walking,
> afternoons sometimes,
> when the joggers were back
> at their desks, along
> San Tomas Creek, as a break

from chairs, computers, and
fluorescent light. The creek,
controlled, mostly cemented,
but for a block or two still running
on a sandy bed, provided
an island habitat for a few
species of plants, animals, and birds.

In the middle of a hundred
square miles of concrete
and asphalt, I took some pleasure
in monkey flowers, alder and willow,
coyote brush, and adventive weeds:
rushes and horsetails, veronica,
water primrose. A few species new
to me I collected, and hung the pressed
specimens in my office. I began to learn
birding: ducks with the seasons,
egrets, black phoebes, kestrel.
A burrowing owl lived there
in a den, killdeer in the shallows,
while a young red-tailed hawk soared
leisurely, overhead.

For some weeks was too busy
to get out, and when did,
was struck by the quiet:
no ground squirrels scurried off
at my approach, not a single killdeer
peeped, nothing moved. It was stilled.
Then I saw: every leaf and branch
of the grasses, shrubs, and trees
was browned and crisping,
and I thought, "sprayed!"

. . . .

Jokingly, I used to tell botanical friends
that part of my work was a study
of the ecology of San Tomas de Aquino Creek.
A foreman
from the Santa Clara Valley Water District
and a busload of boys from the County Farm
cleaned it out with rakes and chainsaws
in three days.
This is my report.

Stories such as these have the capacity to generate deep feelings. But they are not new or news; a single tree being killed, a small section of scrub sprayed in an otherwise controlled waterway. Yet each story captures a wound opened in the Earth and in human hearts bonded with a particular place. And somewhere in each of us, because we sometime have had a similar experience, the buried feelings rise to the surface like bubbles coming to the top of a pond. They break open as thoughts, as feelings. We are, in fact, surrounded by this wound daily, we swim in its waters, and a certain amount of our energy goes each day to repress the feelings that it naturally engenders.

The interior wound occurs in the landscape of the human psyche and heart; I learned how to talk about it in New York City.

New York Stories

Until my first visit, New York City had always lurked forbidding and grim in my imagination. Like most westerners, I knew there were few colors; that the buildings, clothes, and emotional interiors of the people were shades of gray; that the city cowered under sullen storm clouds. That no one ever smiled.

I *knew* what happened to visitors, the harm that came to outsiders. By some mysterious process residents were able to survive the daily beatings, muggings, and murders. And I knew that people from out of town, especially west of the Mississippi, had no such protection. I knew that the inhabitants of the city would be able to tell, immediately, that here was new meat. I was cautious, then, when asked to speak there.

I flew into Washington, D.C., where I had other business, and nervously took a train up to the city. No one met me; my friend was late. So I stood, tiny and dwarfed by the station, bags circled, waiting. My nervousness increased each minute; my peripheral vision expanded completely around my head. My hands casually hovered around my wallet and my eyes constantly roved the crowd, making sure that when the inevitable stranger bumped into me his accomplice did not make off with my bags. By the time my friend arrived I was *very* nervous—she realized this in a glance, laughed, and took my arm. I earnestly shared my thoughts. She patted my arm and reassured me.

As we stepped outside the doors of the station I was astonished; the sky was blue, the people dressed in bright colors, many were smiling. One or two nodded companionably; I began to think I had been misinformed. Leaving me for a moment, my friend strode to the curb, raised her hand; a cab pulled over and stopped.

The cab was delightfully battered, the driver American, not recently shaven, his eye cynical, his clothes disreputable. He loaded the bags as we climbed in the backseat.

"Where to?" His accent was marvelous.

My friend told him, adding conspiratorially, "This is my friend's first visit to New York."

The cabbie turned his head, looked at me. "Welcome to our lovely town," he remarked. "Would you like beer, wine, pop? Dis," he surveyed the interior of the cab proudly, "is a full-service cab."

"But, he's so nice!" I involuntarily exclaimed.

"See! See!" the driver said to no one in particular and, at a high rate of speed, he pulled out from the curb.

I had been invited to New York to give a talk on Sacred Plant Medicine. It had been arranged by someone I knew only slightly and was being promoted by a local organization that would forward my fees later. I was to speak about the process whereby historical indigenous people developed their knowledge of plant medicines and, to some extent, how, in this present time, we could explore that process for ourselves. Over the weeks leading up to the event I had received periodic calls to update me. There were 8 people coming, then 14, then

25—could I come back often? perhaps once a month? there were so many. I was excited. A huge crowd for my first talk in New York City.

Our destination had, like many businesses I had seen on the way, a long green awning stretching toward the street, a few short steps leading up, and a set of impressive doors. Like many of the buildings in New York it had apartments above and a business at street level. I gathered my things, paid the driver—he had actually had no beer, wine, or pop—and we walked up the steps. I grabbed the beautiful brass handles and pulled open the doors.

The air whooshed out and I got a good whiff of the place. The smell was 1950: old creaking offices, ancient elementary schools, empty and echoing government buildings. As my eyes adjusted I saw that the floor was covered with actual linoleum (made from material impregnated with linseed oil) in a large green and white checked pattern. The walls were painted with the dirty-tan, army-surplus paint common on the walls and ceilings of my elementary and high schools. Across the room was an ancient countertop covered with the first Formica, and a hallway behind. No one was present; a few fluorescent lights spluttered and hummed overhead. I felt a sense of gloom.

I peeked around the counter. "Hello?" I called out. Silence. I glanced back at my friend, hesitant. She smiled encouragingly. Tentatively, I made a foray into the darkened corridor. An office on my right. I peered in. Someone was lounging behind the desk, reading a book. A youngish man with upthrust, short-cut hair, an athlete's stocky body. "Yes?" he said.

"Er, I am here to teach the workshop. Where do I go? Can you help me?"

He shrugged. "Everyone is gone, I'm just helping out. Didn't you see a sign?"

"Er, no," I replied. "Uh, you don't know how many people are coming by any chance do you?"

"No," he said, irritated, "but there's a sign out there."

I paused, indecisive. He began reading his book again. I retraced my steps.

"What's up?" My friend asked.

"Er, I don't know, no one seems to know anything. The guy back there said there was a sign out here someplace."

I scanned the room. Nothing. She nudged my arm. "How about that?" I looked closer. Across the room, taped to the wall, was a sheet of lined paper roughly torn from a notebook. It drew me closer. PLANT WORKSHOP was crudely scrawled on it in pencil, an arrow underneath pointed to a shadowy doorway on our left. I walked over, peered in. Steep steps, covered with the dark brown rubberish material that I remembered from childhood, climbed a narrow stairwell. Blue powder was sprinkled along the baseboards.

My friend stuck her head in, "Hmmmm." She looked at me questioningly. My hand clutched the narrow round banister and I started up. The stairs creaked. At the top, under a dim recessed light fixture, was a landing and, on the other side of it, a door. Its twin had filled the doorframe of my high school English class. It was painted the same tan-colored, army-surplus paint and in the upper middle was a small square window of glass. Filled with a lattice of tough metal wire, this glass has always reminded me unpleasantly of juvenile detention. I went in.

Scattered around the room were the tan metal desks from my high school. The battered teacher's desk rested at a rakish angle in front of a blackboard dusty with powdered chalk. Four or five abandoned notebook pages lay scattered on the floor. Light bleakly filtered in through elderly windows to my left. Some desiccated plants in green plastic pots sat along the sills.

Sitting in the front row facing the desk was one young Caucasian woman, about twenty-eight years old. She turned hopefully at our entrance.

I paused in midstride. "Uh, hi . . . uh, uhmmm, I . . . I'll be right back," I stuttered. I set down my bags and raced down the stairs, found the young man still reading.

"There's only one person up there! Is that *all* there is coming?"

He looked up, expression blank. "Oh, I'm sure there will be more in a few minutes." He looked back down at his reading. I stood, hesitant, waiting for him to say more. He kept reading. I reluctantly retraced my steps.

The young woman turned again at my entrance. "Are you the teacher?" she asked.

"Ummm, yes, but there seems to be a problem. You appear to be the only one coming. I think you should get your money back."

It is very hard to teach a class with fewer than five people. There is so little energy that the whole thing never gets off the ground. With one person it's impossible.

Normally a speaker's gaze roves the room, looking at this person and that. Eye contact only occurs with each person every so often. Facial expressions that indicate interest or intent listening or an emotional response to something being said are spaced with time for resting in between. But with only one participant the undiluted gaze of the speaker is trained on her the whole time. In short order she runs through all the appropriate facial expressions and a glaze sets in as the face fossilizes into one, set expression, generally an expression she thinks mimics polite, interested listening. It is actually horribly doll-like—glassy eyes surrounded by unmoving, semi-lifelike, plasticized skin. In short, I was determined, for both our sakes, to avoid it if at all possible.

Then she looked up earnestly and said, "But I've come so far, won't you stay and talk for just a little while?" Something in her expression stopped me and I softened and said, "Yes."

Mentally I gave it fifteen minutes; without giving the talk itself, I thought there couldn't be much to talk about that would take longer than that. And so I sat on the desk, leg swinging, and spoke of how it was for indigenous people long ago, and just as I was wrapping it up the door opened and two more women walked in. This put me over the number of participants where I could, with integrity, cancel the class. Nevertheless, I tried to get out of it.

"Ummm, you know, I think this is all of us. I think you should get your money back."

"Well, hell," one of them said. "We come all the way across town. We ain't got anything better to do. You might as well teach." She put a hand on her hip, daring me to disagree. The other one nodded approvingly. So I waited as they got settled, ranging themselves on either side of the young woman in front of me.

This was strange. People invariably sit scattered around a room; they feel safer to the back or the side—safer still if there is space between them and the person next to them. It is rare that people sit right up front, much less that *all* the participants sit together, in a row, in front.

We were chatting as they settled themselves and something in what they said struck me as odd, so I asked them if they knew anything about herbalism or plant medicines. None of them did. And this was very strange because the people who come to hear me are usually herbalists. So I asked them, "Why are you here?"

And they told me the truth.

The woman on my right, perhaps forty-five years old, answered first. She had been born in Jamaica and her accent was still strong in spite of thirty years in New York. I love Jamaican accents and can listen to them all day long. No matter what they say it sounds beautiful and moving and deeply poetic. ("Yeah, mahn, I goin' to Jamaica mahn, get some rum, den tings be ahree.")

She looked up at me and cleared her throat and there was a funny feeling in the room. She said, "Well mahn, lately I haven't wanted to live." And here she paused and took a deep breath. "But my granmamma, she come to me in my dreams an' she tell me, 'Chile! You got to get your fingers in de dirt.'" She moistened her lips, her voice dropped to a whisper. "Yeah mahn, every night she come to me an' she tell me de same ting, 'Chile! You got to get your fingers in de dirt.'" She shifted in her chair and looked up, going on more briskly.

"So, you know I kep' tinking wha' to do? And den I hear 'bout dis place, a community garten dey call it. It cos' twenty-five dollar. So I save my money for two mont's and take de subway. An' you know what mahn? When I go dere and put my fingers in de dirt I know who I am again." She stopped once more, then looked at me earnestly and said slowly, "Such a ting has nebber hoppend to me before. An' I come today 'cause I tink mebbe I'm crazy."

There was a stillness in the room and I realized the rest of us were holding our breath. There was a sudden rustling of clothes and indrawing of breaths and we all shifted in our seats. I turned to the next woman, the young white woman who had originally been present. She looked like she had been raised suburban middle class, maybe in New Jersey. And I asked her, "Why are you here?"

She looked up shyly, nervously, conscious of the women on either side. "Well," she said, "lately, I have been thinking of becoming a naturopath. So not too long ago I flew out to Portland to the naturopathic college there to see if it was something I wanted to do." She paused and moistened her lips, her head tilted slightly down. "They offer a tour to prospective students, you know, where you go through the buildings with a guide and hear from present or former students how great it is. Well, there were ten or fifteen of us and we were stopped in the middle of this hallway. I wasn't paying attention to what the guide was saying, my mind was wandering, thinking about something else, when out of the corner of my eye I caught a glimpse as she opened a door to my right. I turned and looked and it was the room where they keep all the plants, all the herbs they use for medicine. And I could hear each one of the plants crying out to me, talking as clearly as I am talking to you now. And there were hundreds of them." She paused for a moment, then went on. "I came today because I thought, that perhaps, something in your talk could help me understand what had happened. I have been thinking, you know," and here she moistened her lips again and looked nervously around, "that maybe I'm crazy."

We were entranced, listening to her. And when she was done there was a flurry of breathing and shifting, and a funny, sad feeling in the room. I turned to the last woman. She was, like the first, a black woman, middle-aged, but obviously born and bred in New York City. Her features were hard, her face angular, her cheekbones sharp. She had come up hard and it showed in every movement and intonation, in her clothes, in the fierce determination in her face, and in her strong work-roughened hands.

"Tell us why you are here," I said.

She looked up. "Well, my man, he left me. And I haven't known where to go with my life now he's gone. I live in a little efficiency apartment and I get up early in the morning when it's dark. I work all day and when I come home it's dark again. I don't have a yard," she laughed grimly. "This *is* New York, but I like plants and so I grow them inside . . . in pots." And here she paused and in a shy voice she said, "They like it when you talk to them, you know? They grow better." She shook her head. "Anyway, there is this one. It grows all up the wall and across the ceiling. It has big leaves," she gestured with her arms to show me, holding them in a big circle that narrowed down to a point where her hands came together. "Well, when I get up in the morning, the leaf at the end of the plant, up on the ceiling, it's pointing this way." She held her arms like a big pointed leaf again and gestured off to her left. "But when I get home it's pointing this way" and she gestured off to her right. "I thought that maybe it was the sun, but none of the other leaves were doing it. And every morning and evening it was the same. I kept thinking about it and wondering, and then one day I realized that the plant was trying to tell me what direction to take with my life." Here she stopped and cleared her throat. "You know, nothing like this has ever happened to me before. And I came today because I thought that maybe I was crazy."

The room was quiet when she finished. Ripples from what they had brought into the room lapped against the walls, ebbed inside us, slowed, and were still. And in the stillness I looked inside and saw the wound laid down within all of us. The damage to our interior world from the belief that we somehow crash-landed or inexplicably emerged on a ball of rock hurtling around the sun, the only intelligent inhabitants of Earth. The wound that comes from believing we are alone amid dead uncaring nature. And then I took breath and began to share stories of a time when the world was young, when everyone knew that plants were intelligent and could speak to human beings. When it was not crazy to have your granmamma come to you in your dreams and say, "Chile, you got to get your hands in de dirt!" A time when it was different.

Our disconnection from nature and our disavowal of interior depth—of soul—from animals, plants, and landscapes occurs all the time in all of us. But there is more depth in the world than we have come to believe, than we have been taught. Connection with the interior world of nature has been a part of our species' experience for millennia. Contact with it still occurs when we least expect it: in the glance in a loved one's eyes, the shadowed green in an old-growth forest, the primal power in the majestic walk of a bear. Or, unexpectedly, in dreams of our grandmothers or our daily interactions with plants. Since the words to describe this kind of depth are atrophied or no longer present in our language, the experience, when it does extrude itself, is often difficult for people to deal with; they sometimes think they are crazy—crazy and alone, the only intelligent life-form on Earth.

These wounds are the inevitable expression of an epistemological mistake; they are, *by their nature,* evidence of epistemological error. Regrettably, the logical conclusion of that error is now approaching us at a great rate of speed.

I hold to the presupposition that our loss of the sense of aesthetic unity was, quite simply, an epistemological mistake. I believe that that mistake may be more serious than all the minor insanities that characterized those older epistemologies which agreed upon the fundamental unity.

—Gregory Bateson, MIND AND NATURE

EPISTEMOLOGICAL CONFLICT | 3

The living and holistic biosystem that is nature cannot be dissected or resolved into its parts. Once broken down, it dies. Or rather, those who break off a piece of nature lay hold of something that is dead, and, unaware that what they are examining is no longer what they think it to be, claim to understand nature. . . . Because [man] starts off with misconceptions about nature and takes the wrong approach to understanding it, regardless of how rational his thinking, everything winds up all wrong.

—Masanobu Fukuoka, *The Natural Way of Farming*

The World can, in effect, get along without natural resources.

—Robert Solow, 1987 Nobel Prize, Economics

It would not have seemed fanciful a thousand years ago that one's grandmother had come in dreams to help in a time of trouble. It would not have seemed fanciful that the act of putting one's hands in the Earth would bring a woman to a sense of herself. It would not have seemed fanciful to hear plants speak in human voices, or that a plant would care enough about a human being to try and tell her what direction to take with her life. These things would have been accepted as natural, a normal part of life, and a person considered blessed who had received such help and teachings.

That these things are now considered fanciful, superstitious nonsense, or even crazy marks a significant shift in how human beings view the Universe and their place in it. It is now generally assumed that those older ways of understanding the Universe, those older epistemologies, were and are seriously flawed. That, in fact, such experiences and beliefs are not based on any objective reality and that the people who believe them are at best deluded, at worst crazy. It is assumed that our ancestors, who did believe these things, were uneducated, superstitious, unscientific, and hopelessly misinformed about the nature of the Universe.

Like the majority of people in the Western world, as I grew older I pursued formal schooling. Inevitably, as most people are, I was trained to view the older epistemologies of humanity with suspicion. Indeed, when I saw aspects of those older ways of thinking breaking forth in a

---

Once upon a time what took such good care of me was a guardian spirit, and I damn well knew how to pay it appropriate attention. Despite this invisible caring, we prefer to imagine ourselves thrown naked into the world, utterly vulnerable and fundamentally alone. It is easier to accept the story of heroic self-made development than the story that you may well be loved by this guiding providence, that you are needed for what you bring, and that you are sometimes fortuitously helped by it in situations of distress.

—James Hillman, The Soul's Code

person's thought or action I felt that they were confused or irrational. I was training to become a mathematician.

The Interblending of Language and Perception

For me, the numbers we see each day, such as a 2 or a 3, are only the physical marker, the surface as it were, of the essence of "2" or "3." The number 2 has the same relation to "2" that a physical description of you has to the complete totality that *is* you. The identity and character of numbers, the "properties" that underlie their forms (as with so many things), shape how they behave in the world.

Mathematics is an unusual language. Though mathematical phrases *can* be articulated, mathematics cannot be spoken as are other languages. But like all languages, once a person begins to think in it, it shapes *how* things are perceived and how they can be described. Each language contains words or expressions for experiences that are unique to that language and the culture that gave birth to it. In consequence, many linguistic phrases cannot be accurately translated into other languages, only approximated.

Languages form over innumerable human generations and continue to change as new epistemologies are defined as important and incorporated into the culture. Each culture is a unique experiment in the human encounter with the nature of reality, an experiment carried out over extremely long lengths of time. The language each culture develops encodes unique aspects of what they have experienced and found important as a culture. (And the loss of any language represents the loss of unique information about the nature of the Universe that took perhaps a hundred thousand years or more to gather.)

The more unique experiences are described and refined in language, the more concrete and tangible the experience is to those who think in that language. The French have quite a different relationship with food than the Germans, for instance. Again, this process is a long one and is not consciously done. The children who come long after absorb the language as they grow and those unique, linguistically encoded experiences form an unquestioned element of the fabric of their experience of life. They are a part of a specific cultural epistemology or way of knowing reality. In consequence, language literally shapes personal experience of reality.

I began to notice that as I shifted from thinking in English to thinking in mathematics certain things would come into focus, could be seen, and others went out of focus, could not be seen. I was taking on and internalizing, as my studies progressed, the assumptions embedded within mathematics. That process of internalization had specific and immediate effects. My physiology changed, my personality changed, the range of emotional behavior available to me shifted, and how I interpreted other peoples' behaviors and thoughts shifted. Math-

---

One of the seven terms the Seri use for the different stages of ripeness of mesquite pods, this one [Azj] signified the driest stage, when the pods are easiest to grind into fine flour. The Spanish term for mesquite pod, pechita, *lacks that precision.*

—Gary Paul Nabhan, CULTURES OF HABITAT

Language is . . . under assault by those who intend to control others by first subverting the words and metaphors that people would otherwise use to describe their world. . . . The highly technical language of the expert is . . . useful for describing fragments of the world but not for describing how these fit into a coherent whole. But things work as whole systems whether we can describe them or not, whether we perceive it or not. . . . Consequently as our language becomes increasingly artificial, words and metaphors based on intimate knowledge of soils, plants, trees, animals, landscapes, rivers and oceans have declined. . . . Of the roughly 5,000 languages now spoken on Earth, only 150 or so are expected to survive to the year 2100. Language everywhere is being whittled down to conform to the limited objectives of the global economy and homogenized in accord with the shallow imperatives of the "information age." . . . Because we cannot think clearly, about what we cannot say clearly, the first casualty of linguistic incoherence is our ability to think well about many things.

—David Orr, "Verbicide"

ematical language, like all language, shifts personal experience of reality. Behavior shifts in response.

This experience has many implications. With a friend who is fluent in three languages, his former wife reported that when they first traveled to Greece and he began thinking (and speaking) in Greek instead of English his entire physiology and personality changed. So significant was the shift that she felt she did not know him at all; he seemed a stranger. I have watched him shift from Greek to English to French and his personality changes with each language. He reports that when he begins to think in another language the whole world undergoes a subtle but significant shift; how he perceives reality changes. Different ways of thinking and feeling become available in each new language that were not available in the last. Different but real experiences that each culture has found through long historical encounter with the human condition are encoded in those different languages.

The experiences encoded in mathematics, however, are very different from those in other human languages. When thinking in mathematics human bodies, emotions, and interpersonal relationships take on less and less importance while the Universe as an expression of number relationship takes on more and more. There are ecstatic moments of insight and understanding, of course, but life, human or otherwise, has no linguistic place in the language of mathematics.

Through cultural learning, we acquire a particular worldview that becomes the bedrock of our minds: as deep inside us as language, as far beyond conscious thought as the act of breathing, as potent as our involuntary emotional responses.

—Richard Nelson, "Searching for the Lost Arrow"

In 1962 Thomas Kuhn astonished his academic contemporaries by proposing that scientific theories should be looked on not only as dealing with pure objective facts, but rather as systems of belief.

—Richard Milton, SHATTERING THE MYTHS OF DARWINISM

The many, usually unexamined, assumptions woven into the fabric of mathematics—that numbers exist, that they possess meaning, that the language of number is important, that it is neutral and objective, that all things can be described in mathematical terms, that it is more real than other less precise languages—shape what mathematicians can "see" when they think in mathematical language. It is not a language that can express or see the organic process of life, human or otherwise—no birthing or child rearing, or love or need, no caring or bonding. A person thinking in mathematics cannot perceive the thing that passes between a puppy and a human being. It simply does not exist unless one shifts out of mathematics into another language. I realized, as time went by, that any language I used would shape my experience of the world around me.

This realization took on much more importance to me than that of numbers or number theory. I was shocked then, and still am, that so few people within the sciences understand that the mere act of taking on the language of a particular scientific discipline causes such a shift in human perception, action, psychology, and behavior. A huge portion of the work that scientists have asserted to be objective is affected by elements that are not being factored into their research. And if taking on the assumptions of one scientific discipline has such an effect on individuals, what then is the effect if a whole culture, a whole people, internalizes the epistemology of science itself?

Later I came across the work of the mathematician Kurt Goedel. Asked to examine mathematics and refine the principles upon which it is based he determined that they could not be refined, that they are

---

Science's cosmologies say nothing about the soul, and so they say nothing to the soul, about its reason for existence, how it comes to be and where it might be going, and what its tasks could be. . . . Any cosmology that begins on the wrong foot will not only produce lame accounts; it will also lame our love of existence.

—James Hillman, THE SOUL'S CODE

Scientists often explain what they do as an exercise in applying the "scientific method," but many scholars who have examined the practice of scientific research conclude that the method is more rhetoric than reality. . . . A fact can only be understood through the lens of a theoretical framework. This means that competing frameworks may interpret facts in different ways. . . . The idea that facts are theory-laden is standard in social studies of science. . . . Scientists can "push" their arguments by their choice of technical assumptions, through selective use of evidence and results, by their way of referring to alternative arguments and through their treatment of uncertainties. The implication is that choices between scientific theories are not "scientific" in the sense of being purely logical choices made by neutral researchers based on unambiguous evidence and clear criteria.

—Brian Martin, "Scientific proof and the origin of AIDS"

---

unprovable assumptions. And though mathematics follows logically from the underlying principles (and everything works nicely if you accept those principles as true), they cannot themselves be refined . . . *unless* you stand outside the system itself. You cannot use the tools of a system to refine the system whose tools you are using. And of course, at nearly the same time Werner Heisenberg determined that if an observer assumed electrons were waves they would act more and more like particles and that if they were assumed to be particles they would act more and more like waves.

The implications of Goedel's Incompleteness Theorem and Heisenberg's Uncertainty Principle are routinely ignored in most of the sciences. Specifically: the assumptions (the mental perspectives) of the observer change what is being observed; scientific systems such as mathematics are based on unprovable, often unrefinable, assumptions; and to understand the limits of a system and refine its underlying assumptions it is necessary to stand outside it, to literally be in a different system. But, since science is insisted to be—by scientists, Western culture, public school curricula, and so forth—the only accurate system

Using the mechanistic, reductionist approach of Victorian science, biology has not so much explained life as explained it away. The body is a machine, a matter of chemistry and electricity. Thought is merely a by-product of the computerlike brain which pulls the body's levers. Evolution is no more than a marriage of chance and chemistry. There is no ghost in the machine: human is the machine. It is out of this Frankenstein approach that neo-Darwinism was born and is sustained: by the science of Mendel and Kelvin, rather than of Heisenberg and Planck.

—Richard Milton, SHATTERING THE MYTHS OF DARWINISM

Biota, biosphere, life, and Gaia. These are words whose meaning is imprecise even to scientists; yet they use them with authority. Quotations from sacred books are held in a similar way to be true by the fundamentally religious, but the fundamentalists can argue that for them truth and authority are a matter of faith. Scientists have no such defense for their ignorance.

—James Lovelock, HEALING GAIA

---

through which to view the workings of the Universe, no other legitimate system exists in the Western world in which a person can stand to understand the limits and refine the underlying assumptions of science.

The Epistemologies of Ancient and Nonindustrial Cultures

I was now much more interested in how thinking affects behavior so I shifted my major to transcultural epistemology, the study of how different cultures know. Part of my learning was concerned with the epistemologies of historical and present-day nonindustrial cultures. One observation that has particular relevance for this book stood out strongly: Among widely diverse nonindustrial cultures the members whose specialty was plant medicines, *vegetalistas,* described their experiences remarkably similarly *irrespective of culture, continent, or time.* The vast majority (essentially all instances where I have found first-hand accounts) told interviewers that they did not obtain their knowledge of plant medicines from the exercise of reason or through trial and error. They were uniformly consistent in saying that their personal and cultural knowledge of the medicinal actions of plants came from "nonordinary" experiences, specifically: dreams, visions, direct communications from the plant, or sacred beings. I found this uniformity astonishing. The majority of interviewers were also remarkably uniform, in nearly every case: After being told the source of a practitioner's plant knowledge, they would immediately denigrate it. It would be ascribed to superstition, or ignorance, or un-Christian barbarism. A very few researchers approached their work without prejudice and simply reported *verbatim* what they were told. I found a single instance of a researcher recognizing a pattern, but he did not pursue the implications, merely noted that it was "amazing."

Nearly all scientists insist that indigenous peoples learned the uses of plants through a lengthy trial-and-error process. There is an immediate problem with this assertion, of course; they were not there to observe it. Their assertion is an assumption, a guess, though so widely repeated it has taken on the mantle of fact. But let's look at how that assertion might play out in the real world.

Yarrow *(Achillea millefolium)* is a pervasive plant throughout the Northern Hemisphere. Historically, nearly everywhere yarrow grows, it has been used as a hemostatic herb: to stop bleeding, especially bleeding caused by wounds. A great many of its traditional names indicate this property: The ancient Roman name *herba militaris* (soldier's grass) and the North American Teton Dakota tribal designation *tao-pi pezu'ta* (medicine for the wounded) being two. The Latin name itself, *Achillea millefolium,* means "the thousand-leaved plant of Achilles"—who used it to heal men hurt in battle. Even its common name, yarrow, is old English for "spear well"—to make well from spears.

This knowledge of the use of the plant is assumed by scientists to have occurred through trial and error. From a superficial examination this assertion seems to make sense. Each of us has learned things through trial and error: Never cut toward yourself with a sharp knife, for instance. But let's imagine how this might have occurred with yarrow.

Imagine a man a very long time ago in a forest (the first man ever to do this, in fact), and he cuts himself and begins to bleed. The cut is a bad one—lots of blood. He decides to put a plant on the wound. I have often wondered at this; it seems a radical decision. Still, he begins to place plants on the wound. He tries some grass. Nothing. Marsh mallow leaves. Nothing. The bleeding is getting worse so he is moving through the forest with greater speed trying this and that. Finally he grabs some yarrow and places it on the wound (first, of course, bruising it so that the juice of the plant liberally enters the wound). The bleeding stops. Yarrow stops bleeding. He tells everyone in his immediate area and knowledge of this plant medicine enters his cultural lore. By a similar process a member of every other culture on Earth makes a similar determination and yarrow enters *their* cultural lore as well. As far as this goes it can seem to make sense. But let's extend this exercise a little—to plants of the Artemisia family.

Rather than applying the epistemology of Western understanding to the alien, the tribal, and nontechnological cultures we [s]hould let their anthropology (their stories of human nature) be applied to ours.

—James Hillman, THE SOUL'S CODE

Throughout the world, in addition to many other medicinal uses, artemisias are used to ward off negative influences, bad energy. As an example: Melvin Gilmore quotes a Dakota Indian of North America regarding *Artemisia ludoviciana* that it serves as a "protection against maleficent powers; therefore it was always proper to begin any ceremonial by using Artemisia in order to drive away evil influences."[1]

So, here we have a man in the forest again, perhaps just walking along minding his own business, and he encounters negative influences, maleficent powers. He is quite afraid (as anyone would be) and begins looking for a plant to help. He holds yarrow up toward the negative influence but it fails to work. Grass, nothing. Marsh mallow, no. Cherry leaves—no effect. Panicking now, he rushes through the forest trying more and more plants until, finally at the last minute, he grabs an *Artemisia* and holds it up toward the negative influence. The negative influence dissipates. *Artemisia* wards off negative influences. He shares the information and this use of *Artemisia* enters his people's cultural lore. By a similar process it enters the knowledge base of all other cultures on Earth.

For trial and error to be the method by which this plant information was gained (and not visions or dreams or talking plants), it must be assumed that negative influences, maleficent powers, exist. Most scientists will not concede they exist, much less that they can be perceived, still less that a plant can ward them off. Yet throughout the world the vast majority of cultures identified this action of *Artemisia* uniformly.

Botanical researcher William Chase Stevens comments:

> In the New World, as in the Old, the lives of the natives were intimately and vitally related to the plant population, and it need not surprise us that our Indians put the indigenous Artemisias to much the same medicinal uses as the early Europeans and Asiatics did theirs; but that our Indians should have, as they did, the same kind of superstitions about the Artemisias and use them in similar rites and ceremonies, with confidence in their magic powers is amazing.[2]

Stevens acknowledges, which I have seen no one else do, the startling symmetry of thought among cultures geographically and temporally isolated about the transcendent aspects of *Artemisia*. Second, and less honorably, he blithely describes these beliefs as superstition (in "our" Indians). This kind of classifying of the commentary of nonindustrial cultures about the source of their plant knowledge represents a bias on the part of the reporter/observer that comes from unacknowledged internalized assumptions about the nature of reality. The comment says more about Stevens and his culture than it does about Indians.

Indigenous peoples were clear, however, about where their knowledge of plant medicines originated. In the vast preponderance of cases, when they were asked, they insisted that their knowledge of plants came, not from trial and error, but from the plants themselves, from visions or dreams or from sacred beings. That their description of the sources of plant knowledge should be so uniform is in itself, as Stevens notes, amazing. The assumption by scientists that all nonindustrial peoples generated these descriptions out of superstition and ignorance is astonishingly shortsighted and, frankly, not very good science. What is especially striking is that the medicinal uses for plants that nonindustrial people were taught during these experiences correspond nearly perfectly to the medicinal actions of the plants that have been identified through science.

As a friend told me many years ago on a particularly bad day when I had no desire to self-reflect: "If one person calls you a horse's ass, he might be wrong. If everybody starts to, it's time to saddle up." When presented with such a uniformity of information from disparate cultures separated so widely by geography and time, any decent scientist

---

[In my vision] we were facing the east and I noticed something queer and found out that it was two men coming from the east and they had wings. On each one's breast was a bright star. . . . They had an herb in their hands and they gave it to me, saying, "Behold this; with this on earth you shall undertake anything and accomplish it." As they presented the herb to me they told me to drop it on earth and when it hit the earth it took root and flowered. You could see a ray of light coming up from the flower, reaching the heavens, and all the creatures of the universe saw this light.

—Black Elk (in DeMaille, THE SIXTH GRANDFATHER)

In infinite detail her internal organs appeared on the screen of my vision. As the liver came into my sight, it was obvious from its black color that it had ceased to function and I knew it was no longer serving to purify the blood. As this became clear to me I turned my attention to the remedy, and the appropriate plants appeared in my vision—flowers from the retama *tree and roots from the* retamilla *shrub. As the visions faded off into more general dreams I knew it was possible for her to recover [which she did after I had treated her].*

—Manual Cordova Rios (in Lamb, RIO TIGRE AND BEYOND)

Invisibility perplexes American common sense and American psychology, which hold as a major governing principle that whatever exists, exists in some quantity and therefore can be measured. . . . When the searchers failed to find the soul in the places where they were looking, scientistic psychology also gave up on the idea of the soul.

—James Hillman, THE SOUL'S CODE

---

would have saddled up and been ready to ride a long time ago. Unfortunately, scientific epistemology has historically made it impossible for most scientists to "see" what indigenous people were saying. But what if nonindustrial cultures were not engaging in superstition but describing an actual event? This would mean that there is another way to gather information about the Universe and our place in it than what we call "science." (It may be helpful here to remember that the realization of the double helix structure of DNA came to its discoverer not through scientific study but in a dreamlike state, whole and complete.)

Embedded within the underlying epistemologies of the vast majority of nonindustrial cultures are the components of this other way of gathering information. While containing numerous variations, themes, and differences these nonindustrial epistemologies do contain a basic framework that is very similar in a number of areas. Most assert that:

- At the center of all things is spirit. In other words, there is a central underlying unifying force in the Universe that is sacred.
- All matter is made from this substance. In other words, the sacred manifests itself in physical form.
- Because all matter is made from the sacred, all things possess a soul, a sacred intelligence or *logos*.
- Because human beings are generated out of this same substance it is possible for human beings to communicate with the soul or intelligence in plants and all other matter and for those intelligences to communicate with human beings.
- Human beings emerged later on Earth and are the offspring of

the plants. Because we are their offspring, their children, plants will help us whenever we are in need if we ask them.

- Human beings were ignorant when they arrived here and the powers of Earth and the various intelligences in all things began to teach them how to be human. This is still true. It is not possible for new generations to become human without this communication or teaching from the natural world.
- Parts of Earth can manifest more or less sacredness, just like human beings. A human being can never know when some part of Earth might begin expressing deep levels of sacredness or begin talking to him. Therefore it is important to cultivate attentiveness of mind.
- Human beings are only one of the many life-forms of Earth, neither more nor less important than the others. Failure to remember this can be catastrophic for individuals, nations, and peoples. The other life in the Universe can and will become vengeful if treated with disrespect by human beings.

This outline, in a very rough way, represents, perhaps, the oldest epistemology of humankind and was present in most historical cultures on Earth. Cultures codified this in differing ways, described it in different words, rigidified it as religion in varying forms. But beyond its varying forms of expression in different cultures and beyond its classification as religion it represents a way of describing human relationship with the Universe and Earth. If all the emotional connotations are removed this epistemology is not so different from some of the descriptions that scientists have of the world, the Universe, and human beings. (For example: $E=MC^2$; green plants gave rise to the oxygen atmosphere that allowed the evolution, the birth, of human beings; by studying the Universe we can gain knowledge and not be ignorant of the things around us; failure to live in an ecologically balanced way can and probably will lead to environmental disaster.) There are, as well, a number of major differences. Though historical indigenous and nonindustrial peoples were exceptionally gifted empirical observers, they did not believe that the Universe is a great machine which, by disassembly,

From Wakan Tanka, the Great Spirit, there came a great unifying force that flowed in and through all things—the flowers of the plains, blowing wind, rocks, trees, birds, animals—and was the same force that had been breathed into the first man. Thus all things were kindred, and were brought together by the same Great Mystery.

—Luther Standing Bear (in Nerburn, NATIVE AMERICAN WISDOM)

---

examination, theoretical exploration, analysis, and study can be understood, its workings made plain. They ascribed to the Universe livingness, interior depth, intelligence, soul, and a central unifying spirit at its core.

TURNING THE UNIVERSE INTO A MACHINE

The journey from a sacred intelligence at the center of the Universe to a generally accepted belief that the Universe is a great machine devoid of spirit occurred over a number of centuries, perhaps as few as five or six, perhaps as many as twenty. It has accelerated greatly during the past seventy-five years. This major shift in how the Universe is viewed is exceptionally new in geologic time. It received great boosts during the Renaissance—from Nicholas Copernicus (1473–1543), Galileo (1564–1642), Johann Kepler (1571–1630), and Francis Bacon (1561–1626). But it was raised to its primary importance by René Descartes, the French mathematician and philosopher (1596–1650). And as it spread in popularity, this newly emerging epistemology of reductionism began to shape how reality was, or could be, perceived.

Prior to the rise of Christianity, the Romans believed, like most historical cultures, that sacredness and intelligence were present in everything. As a response (in part) against Roman paganism, Christianity, as it attained political power during the fall of the Roman Empire, began a process of narrowing other religious expressions and theologically removing sacred intelligence from everything except the Trinity and those deities (such as angels and saints) designated by the Church to be sacred. The Church also inculcated the biblical imperative to exert dominion over nature, a nature created expressly for people by God, which, in and of itself, made nature more of a "resource" and less sacred than the Romans held it to be. And some of the fears of nature's dark side, of disease, and unavoidable fate, and of the wild, powerful, unpredictable Nature of the Roman religion—seeped in as well. Nature not only was a resource but something to be *controlled* and ordered for Man's use. The Wild redeemer abandoned for the Light. Protestantism carried this further, abandoning saints and angels, and reducing sacredness in physical form to a single expression: Jesus (and, possibly, the Bible). Modern (scientific) thinkers, beginning during the Renaissance, took this logic even further: there was no sacredness in the Universe at all—it was, and is, just a machine. Or as David Ehrenfeld comments:

> It only remained to diminish the role of God, and we arrived at full-fledged humanism. . . . The transition to humanism was an easy one; it could occur in steps. One only had to start with the belief that humans were created in God's image. God could then be retired on half-pension, still trotted out at the appropriate ceremonies, wearing the old medals, until bit by bit He was demystified, emasculated, and abandoned. The music that accompanied this process, in its later years, was the throbbing of Watt's steam engine. "Here," it pulsed, "is real power, power, power." To this, the advocates of traditional religion found no satisfactory answer. . . . Had they not, after all, created this godless monster, humanism, with all their endless chatter about our inheritance and dominion over the earth? What did they expect?[3]

Perhaps no one took this concept further, nor influenced modern thinking more, than René Descartes. Descartes insisted the universe was a mathematical system of matter in motion, mechanical in nature. As Rupert Sheldrake comments, Descartes "applied this new mechanical way of thinking to everything, even plants, animals, and man. . . . He laid the foundations for the mechanistic worldview in both physics and

The category error of the neo-darwinists is like that of René Descartes (1596–1650), who distinguished humans from all other living things in alone possessing a soul. Both Descartes and Richard Dawkins limit a general property, soul or life, to a specific class of owner.

—James Lovelock, HEALING GAIA

Alienated from nature, human existence becomes a void, the wellspring of life and spiritual growth gone utterly dry. Man grows ever more ill and weary in the midst of his curious civilization that is but a struggle over a tiny bit of time and space.

—Masanobu Fukuoka, THE NATURAL WAY OF FARMING

In Western society, we rest comfortably on our inherited truths about the nature of nature. Our burgeoning environmental literature, for example, contains a nearly endless variety of statements about the absence of mind in nature. The environment is numb to a human presence—blind, deaf, inert, insentient, compassionless, sometimes brutal in its raw, random power. . . . Despite our certainty on this matter, the anthropological literature indicates that most of humankind has concluded otherwise.

—Richard Nelson, "Searching for the Lost Arrow"

---

biology. In the philosophy of Descartes, souls were withdrawn from the whole of the natural world; all nature was inanimate, soulless, dead rather than alive."[4]

There are, however, implications to the proposition that all things are part of a great machine, that they are neutral in value, and without essential soul or interior depth. If *all* things are part of a great machine, merely parts in its functioning, then human beings are merely parts as well, without essential meaning, only machinelike bits. Accepting this kind of thinking, internalizing it, gives rise to a sense of existential despair, a feeling that one's personal life is meaningless.

Descartes's response was his famous assertion *Cogito ergo sum,* I think therefore I am. In other words, if you think, you have meaning, interior depth, "amness." You are outside the restrictions on the bits that make up the Universal machine. *Cogito ergo sum* also asserts its opposite: if you do not think, you are not. (In many ways this formulated our present dichotomy between thinking and feeling—those who are given to feelings and intuitions or ungoverned passions are not thinking, therefore their level of beingness, their value, is less than those who do think.) These are significant assertions and are essential to scientific epistemology. For the first time in the history of human habitation of Earth value as a being became inextricably attached to thinking. The more something (or someone) thinks the more value or interior depth it has. And the less something (or someone) thinks the less value it has—the more it is simply a thing, a thing that can be utilized by those *with* value, those who think. Culturally we are embedded in this view more deeply than we realize; we are taught it from the moment we are born, our school systems, from kindergarten through college, are based in it. Thus, among all things in the Universe *people* think the most and so have the most value. But even among people, value ("amness") rises or falls depending on how much or well one thinks: Culturally a physicist has much more value than a janitor. And, of course, we think better now than people a hundred years ago and so possess more worth. We "know" as well that people a hundred years from now will know more than us and that, in some strange way, as a result, will have more value. We all know how this hierarchy has traditionally looked: civilized, educated white men at the top. Then, in descending order: white women, white children, people of color (i.e., servants), indigenous people (i.e., savages), then either dolphins or chimpanzees (scientists are not sure which), fur-bearing animals, other sea-living mammals, and so on down the ladder of value—reptiles, plants, algae, bacteria, and stones (nothing is dumber than a rock, except maybe air molecules).

This epistemology is of course under strong attack by women, people of color, and indigenous peoples. Their desire is to have all people equally at the top—given equal value as thinking beings. And great strides have been made in this direction. Nevertheless, from there the descent from more to less value is pretty much the same, from children on down.

At the same time that Descartes was developing his perspectives of a mechanistic universe, Francis Bacon was developing his perspectives about how that universe should be treated and by whom. Sheldrake comments that:

> [Bacon] proclaimed that nature "exhibits herself more clearly under the trials and vexations of art [mechanical devices] than when left to herself." In the inquisition of truth . . . nature was to be "bound into service" and made a "slave" and "put into constraint." She would be "dissected," and by mechanical arts and the hand of man, she could be "forced out of her natural state and squeezed and moulded," so that "human knowledge and human power meet as one." He advised the new class of natural philosophers to follow the model of miners and smiths in their interrogation and alteration of nature, "the one searching into the bowels of nature, the other shaping nature as on an anvil." And he wrote of the new science as a "masculine birth" that will issue in a "blessed race of Heroes and Supermen."[5]

These heroes and supermen would form the basis of "a technocratic utopia, in which a scientific priesthood made decisions for the good of the state as a whole."[6] Their rationalism and technology would, as David Ehrenfeld comments in *The Arrogance of Humanism,* "protect us from the Darker side of Nature . . . winds, frosts, droughts, floods, heat waves, pests, infertile soils, venoms, diseases, accidents, and

---

When we ask "what is life?", we are shown a gamut of living organisms. Mammals first, of course, for toads and frogs seem less alive, and trees and plants less still, and lichens, algae, and soil bacteria, hardly alive at all. Much of the instinctive objection to viewing the Earth as a living system comes from our zoocentrism, the tendency to consider ourselves, and animals, as more alive than other living organisms.

—James Lovelock, HEALING GAIA

[the] general uncertainty that it offers in succession or simultaneously."[7]

Because nature and its parts are only a machine and parts of a machine, anything can be done to them in the search for their inner workings. "Some of Descartes's followers explicitly denied that animals could feel pain; the cry of a beaten dog no more proved that it suffered than the sound of an organ proved that the instrument felt pain when struck."[8] Descartes engaged in exhaustive research on live animals. He himself noted that

> if you slice off the pointed end of the heart in a live dog, and insert a finger into one of the cavities, you will feel unmistakably that every time the heart gets shorter, it presses the finger, and every time it gets longer it stops pressing it.[9]

In short, there is only humankind on the one hand and a mechanical universe, which may be utilized in any way people see fit, on the other. Charles Darwin's (1809–1882) theories about the origin of species through descent by natural selection were generated out of the mechanistic epistemology formulated by Descartes and others. The ascent of man (as Jacob Bronowski called it) through evolutionary development and natural selection that is commonly taught in schools is amazingly similar to the hierarchy of thinking that evolved out of Descartes's assertions about thinking and being.

The formulation of universe-as-machine, the development of the scientific method to study it, the formation of schools to teach the method, the discoveries made about the workings of the Universe after

---

Do you really believe that humans invented the wheel out of their big brains alone, or fire, or baskets, or tools? Stones rolled downhill; bolts of fire shot from the sky and out of the earth; birds wove and probed and pounded, as did apes and elephants.

—James Hillman, THE SOUL'S CODE

those assumptions were accepted, are all heralded as the dawn of the golden age of Man: the beginning of the scientific age, the beginning of the development of sophisticated technology, the move by mankind away from superstition to a way of viewing the Universe based on fact.

Limits of the Epistemology of Science

What is interesting to me is what happens to human behavior, individually and culturally, when differing epistemologies are taken on and internalized. The more widely that science is internalized by cultures and individuals, the more clearly they are significantly affected by such internalization. This leads to questions: Does the wide acceptance of the epistemology of science, specifically seeing universe-as-machine, support successful habitation of Earth? Does it support healthy habitation inside a human life for individual human beings? Does it maintain the health of cultures?

Science is really a remarkable human invention. But in all the excitement of its usefulness the fact that it *is* an invention is often forgotten. As with all human inventions it possesses limitations, unexamined assumptions, and design flaws.

> *This is the weakness of our century—an uncritical acceptance of technological breakthroughs as bounties free from harmful side-effects.*
>
> —Marc Lappé, When Antibiotics Fail

Because of the strong conflicts between fundamentalist Scientists and fundamentalist Christians, it is very difficult to examine the limits of science or standard evolutionary theory without being labeled a fundamentalist Christian or unscientific, or even as someone who hates science itself. That there are a diversity of perspectives about science among many different kinds of people is often overlooked. Many scientists and the general public are often unwilling—indeed, will antagonistically refuse—to openly explore problems in the structure of science and their possible ramifications. This unwillingness by so many scientists to self-examine their profession is exceedingly disturbing.

Perhaps science should be more like knives: understood to be a useful tool that, if misused, can be dangerous. Knives can accidentally

Neo-Darwinism is seen by many of its adherents as the citadel of rationalism against the incursion of the barbarians of unscientific New-Age thinking. One unexpected result of this fanatical defense is that scientific rationalism, which used to be a badge of honor and a beacon of hope for the future, has sometimes become the white sheet and hood of bigoted closed-minded thinkers. And what I find particularly fascinating about this kind of thinking is that it is pretty nearly the exact opposite of the truth.

—Richard Milton, SHATTERING THE MYTHS OF DARWINISM

We found it, in fact, impossible to publish Gaia papers in such learned journals as Science *or* Nature. *There was no lack of interest in Gaia; there was just an apparent determination by peer reviewers (the panels of science arbiters who vet items for acceptance) to deny publication to the idea; indeed, they seemed to regard it as dangerous.*

—James Lovelock, HEALING GAIA

---

wound, or even, in the hands of the deranged, be used to kill. Yet individuals and cultures understand the potential danger of knives and allow for it in their daily use. Knives, unlike guns, are not assumed to be inherently evil in contemporary debates nor, like science, to be always beneficial and harmless.

How science may cut us as a species if we let our attention wander or if a madman gains control of it is not well understood in our culture, and any examination of this aspect of science is often strongly discouraged. This prevents science from being as well integrated into human cultural life as a knife—a useful tool whose strengths and weaknesses are well understood. Because science is assumed to be essentially beneficial with no shadow side it is impossible to know how to use it safely—by cultures or individuals.

Though a great many technological advances have come from the shift to the newer epistemology of science, growing evidence from a variety of fields (such as the study of antibiotic-resistant bacteria) shows that a number of assumptions upon which the shift is based are, in fact, questionable. The assumptions underlying science allow the Universe to be broken down into smaller and smaller parts and permit those parts to be studied in exquisite detail. This analysis of the parts and their interactions with each other has led to the majority of technological advances we now enjoy—science's major strength. The weaknesses of these assumptions are only now becoming apparent. Insidiously: It is not possible for the majority of scientists to put the parts back together and grasp the intricate interweaving of the whole. James Lovelock comments on this limitation of reductionism in his book *Healing Gaia*.

> The problem with reductionism lies with the belief that the method of examining systems by taking them apart is all that is needed. Reductionists are certain that there is nothing in the whole system that cannot be predicted from a knowledge of the parts. . . . [But] To understand [the universe] and its most complex entities—living systems—reduction alone is not enough.[10]

Reductionism presents further significant problems. The greatest indicator of fundamental errors in the epistemology of science is what happens over time to the people and nations that internalize it as a primary epistemology. Specifically: How do they treat other people, other life-forms, the environment? What happens to their culture?

Historically, the internalized assumptions of science have led to the denial of equal interior depth or intrinsic value to women, people of color, and indigenous societies and peoples. They thought differently (i.e., less) and, as a result, they were assigned less value and it was assumed that their interior worlds were less deep (much as a rock's interior world is considered to have less depth). The arguments for denial of equal value to these categories of people were primarily developed by scientists (though many Christians utilized religious doctrine to support it as well).

Utilizing the most sophisticated contemporary scientific thought and studies prominent scientists developed lengthy theories proving blacks were not human but animals intermediate between the great apes and men. They could be owned because they had no soul, no intrinsic depth. They could be (and were) used in medical research as experimental animals without the use of anesthetics.

Dr. J. Marion Sims, considered to be the father of modern gynecology, developed the surgical remedy for vesicovaginal fistulas (an opening between the bladder and the vagina) by operating on the vaginal walls of black slaves—without the use of anesthetics—some of whom he operated on more than thirty times. The fistulas had been created, in some of his subjects, by his own actions.[11]

Darwinist anthropologists utilized extensive studies of the Negroid skull and studies of their earwax to prove that they were not related to the white race; that they were a different order of animal. Richard Milton comments that

> Thomas Huxley . . . observed that "No rational man, cognizant of the facts, believes that the Negro is the equal, still less the superior, of the white man." Darwin himself founded much of his evolutionary thinking on equally racist ideas. In *The Descent of Man* he indicated his belief that the Negro races were more closely related to the apes than white people and also his belief that, "at some future period, not very distant as measured by centuries, the civilized races of man will almost certainly exterminate and replace the savage races throughout the world."[12]

Women were shown to be an inferior form of man and denied the right to education, property ownership, and the vote. Indigenous peoples were considered little better than animals and extensive scientific research "proved" that they could not be educated (civilized), that they were, at best, simple children, and thus could not own the land they lived on because they were not capable of rational thought. This type of thinking and its extension to specific groups of people reached its peak in Nazi Germany.

Germany at that time was *the* most scientifically advanced nation on Earth. The attitudes inculcated about the Jews and other "undesirable" races were developed by the world's leading scientists and came directly out of universe-as-machine and Darwinist perspectives. As Raoul Hilberg, an expert on the Holocaust and Nazi ideology, has said, it is not the common people who begin this way of thinking but the experts.[13] Scientists formulate a theory reducing one class of person to a lesser status, the lawyers and legislators turn it into law, the police carry it out, and it enters popular thought as a "given," as natural law.

The Nazis viewed Slavic peoples with even greater dismay than they did the Jews, killing eleven million of them. German scientists insisted that Slavs were subhuman.

> The subhuman, this apparently fully equal creation of nature, when seen from the biological viewpoint, with hands, feet, and a sort of brain, with eyes and a mouth, nevertheless is quite a different, a dreadful creature, is only an imitation of man with man-resembling features, but inferior to any animal as regards intellect and soul. In its interior, this being is a cruel chaos of wild, unrestricted passions, with a nameless will to destruction, with a most primitive lust, and of unmasked depravity. For not everything is alike that has a human face.[14]

The removal of equal value from Slavs—their reclassification as a not-alive part of the machine of the Universe—was a process extended to Gypsies, the retarded, homosexuals, and Jews. Because they were not human they could be used as experimental animals and German physicians explored a wide range of research: from immersing captured Russian pilots in ice-cold water and precisely recording their physiological symptoms as they slowly died (to gain knowledge to help those suffering from hypothermia), to "shooting bullets into children's legs in order to examine the onset of gangrene and try out treatment methods."[15]

At one time or another, using scientific rationale, women, blacks, Asians, and indigenous peoples all have been denied to be fully human or equal to white men. Christopher Stone, in his law review article

"Should Trees Have Standing?" comments: "The first woman in Wisconsin who thought she might have a right to practice law was told she did not, in the following terms":

> The law of nature destines and qualifies the female sex for the bearing and nurture of children of our race and for the custody of the homes of the world. . . . [A]ll life-long callings of women, inconsistent with these radical and sacred duties of their sex, as is the profession of the law, are departures from the order of nature; and when voluntary, treason against it.[16]

It was argued, using the same reliance on the scientific order of nature, that blacks and Asians were incapable of the same level of rational thought as whites. In consequence they were not reliable in court as witnesses.[17]

Groups who have been disenfranchised by this way of seeing the world, as they have each gained social power, have eventually been accorded equal interior depth and value—though there is still a lingering belief among some that women and blacks do not think as well as white men. The horrors that have come from the denial of value to certain groups by those in power have led to the struggle to expand the zone of value and subsequent legal "standing" to more and more people. It is exceptionally rare, however, for this "standing" or value to be extended to nonhumans—to trees or lizards, to plants or bacteria. Extending equal value to plants—to treat them as human beings—is as laughable and inconceivable to most people now as it once was to extend equal value to women or slaves. Or, as Theodore Roszak put it, "as the prevailing reality principle would have it, nothing could be greater madness than to believe that beast and plant, mountain and river have a 'point of view.'"[18] As a result, trees and lizards and plants and bacteria can be treated as things to be owned, cut down, experimented on, sold, processed, killed, or consumed without regard for any interior depth or intelligence on their part. The exterior wound.

Losing Connection to the Living World

The internal effects on people of the scientific epistemology are more subtle but just as painful as the effects on the rest of nature. Once the Universe becomes a machine, no longer alive, once human beings are defined as the only intelligent life-form, a unique kind of isolation enters human lives, a kind of loneliness that is unprecedented in the history of human habitation of Earth. It is a source of many of the emotional pathologies people struggle with. In addition, people begin to judge themselves internally, to identify their level of value according to how much or how well they think. Any internal expressions, perceptions, or thoughts that come from older epistemologies—that are based primarily on feeling or intuition or aliveness in the Universe—they label unscientific, irrational, unreasoning, or illogical. Such thoughts and perceptions, it is assumed, have less value, are based on improper assumptions about the nature of reality, and are therefore something to be discounted, dismissed, degraded. This dynamic has become so ingrained that people routinely monitor and censor perceptions that are contrary to universe-as-machine. And so people cut themselves off from the Universe in which they live; they become passengers on a ball of semimolten rock hurtling through the Universe. They internally denigrate and deny their most basic experiences of the livingness of the world in which they live, their connection to it, and the importance of that connection. The interior wound.

The tension between these two perspectives—Universe as alive and universe-as-machine—is readily perceivable. Consider Norbert Mayer's poem

Just now
A rock took fright
When it saw me
It escaped
By playing dead[19]

and contrast it with Ken Wilber's observation:

[A]ll you and your pet rock can share is, you fall at the same speed.[20]

If you let yourself relax into the statements and internalize them you will notice a distinct difference in how each of them feel.

I recite Mayer's poem to many hundreds of people each year; the reaction is invariably the same. As the last line is recited their faces light up and they laugh: It reminds each of us of something we have always known—that the Earth, plants, rocks are somehow alive. The realization is accompanied by a natural, childlike joy that is immediately felt throughout the body.

Wilber's statement, in contrast, is quite different in its effects. Sometimes people do laugh, though not nearly so many as with Mayer's poem. The use of the word "pet" as an adjective before "rock" creates a sense of sarcastic denigration, so the laugh is a sarcastic one: at Wilber's cleverness, the thought of a person being dumb enough to think of a rock as a pet, and the image of a person and his friend the pet rock falling together. Most often there is a silence as the sentence penetrates the body, then a sigh of pain, a general sadness. The childlike joy accompanying Mayer's poem dissipates. Mayer's poem activates a sense of personal aliveness and childlike wonder. Wilber's moves the listener up out of the body into the mind, into thinking, into sarcastic cleverness; some listeners feel dumb or foolish. And perhaps this is the point of the comment: To make foolish the belief, and anyone who would espouse it, that rocks are somehow alive, that they have equal value to people, that there is this kind of livingness in the Universe.

The coming ecological disaster we worry about has already occurred, and goes on occurring. It takes place in the accounts of ourselves that separate ourselves from the world.

—James Hillman, THE SOUL'S CODE

Each statement represents a distinct epistemology. As each is internalized they shape individual experience and perception.

Is the soul solid, like iron?
Or is it tender and breakable, like
the wings of a moth in the beak of the owl?
Who has it, and who doesn't?
I keep looking around me.
The face of the moose is as sad
as the face of Jesus.
The swan opens her white wings slowly.
In the fall, the black bear carries leaves into the darkness.
One question leads to another.
Does it have a shape? Like an iceberg?
Like the eye of a hummingbird?
Does it have one lung, like the snake and the scallop?
Why should I have it, and not the anteater
who loves her children?
Why should I have it, and not the camel?
Come to think of it, what about the maple trees?
What about the blue iris?
What about all the little stones, sitting alone in the moonlight?
What about roses, and lemons, and their shining leaves?
What about the grass?

—Mary Oliver, "Some Questions You Might Ask"[21]

In earlier times, when nature was perceived as alive, with intelligence and soul, a natural process took place. People bonded with nature much as people bond with their pets or family now. This bonding process—which has decreased in frequency the more the mechanistic worldview has pervaded society—engendered a certain kind of attitude toward nature. It is an aspect of what Edward O. Wilson calls *biophilia*—a genetically encoded or innate emotional affinity toward all other life-forms on Earth. But the more that children are taught that thinking defines their value, that Earth is dead, that other life-forms intrinsically possess less value, and the more they are separated from regular contact with wild nature, the less biophilia occurs—the less the genetic encoding for caring for and bonding with life is initiated.

Biophilia: bio—life; philia—a tendency toward or an excessive affection or fondness for. From the Greek words bios*—life—and* phileein*—to love. In other words, a deep fondness for, connection to, and love for life forms and living things.*

Biognosis: bio—life; gnosis—knowledge or recognition, especially spiritual knowledge or insight. From the Greek words bio*—life—and* gignoskein*—to know. In other words, to know life through the deeper spiritual and intuitive faculties of the mind; also the body of accumulated knowledge that comes from perceiving life in this manner.*

The Loss of Biophilia and Biognosis | 4

Nature and I are two.

—Woody Allen

Human history did not begin eight or ten thousand years ago with the invention of agriculture and villages. It began hundreds of thousands or millions of years ago with the origin of the genus Homo. For more than 99 percent of human history people have lived in hunter-gatherer bands totally and intimately involved with other organisms. . . . As language and culture expanded, humans also used living organisms of diverse kinds as a principal source of metaphor and myth. In short, the brain evolved in a biocentric world, not a machine-regulated world.

—E. O. Wilson,
"Biophilia and the Conservation Ethic"

Human beings, throughout most of their habitation of Earth, have been so completely interwoven into their environment that, until recently, there was no separation between them. This understanding is reflected in information shared by the majority of indigenous cultures: they did not experience themselves and nature as separate entities. The intimate interweaving of humanity with the rest of life throughout evolution means that the entire development of the human species as a distinct species cannot be separated from the landscapes in which it developed. Such deep interconnectedness to environment is so fundamental to us as a species that, ultimately, it is not possible to understand ourselves as human beings without understanding something of wild nature itself. Because the experience of nature and other life-forms is so deeply interwoven into our emergence as a species, human beings possess a genetic predisposition for wild nature and for other life-forms—though it must, through specific experiences, be activated. Edward Wilson calls this innate feeling or caring for living forms and systems, for nature, *biophilia*. This innate affinity for all life-forms—for things that do not even appear alive to Western perspectives—can be understood even more if the emergence of the human species is seen through the lens of Lynn Margulis's work with serial endosymbiosis.

Our Bacterial Ancestors

The standard evolutionary perspective taught in most schools is that species evolved by a combination of specialization through natural selection (the fastest gazelles surviving to reproduce until all gazelles are fast) and random mutation; this is Neo-Darwinism. Traditional biologists generally view all species as isolated units unconnected to the larger ecosystem within which they live or the larger system of Earth itself. This perspective—that we or any life-form can be viewed in isolation from all other life-forms—has been encoded in our present cultural epistemology. It is a far from accurate view and has had tremendously negative impacts on Earth and human ecosystems.

Lynn Margulis, a codeveloper of the Gaia Hypothesis with James Lovelock, is one of the few researchers who has succeeded in challeng-

Biologists often have only vague knowledge of the nearly four-billion-year history of life on earth. And much of what they think they know is likely to be outdated, oversimplified, or downright wrong: mammals outcompeted dinosaurs, the Pleistocene glaciation produced a mass extinction, evolution is a reasonably predictable progression from simple to complex organisms, and so on.

—David Raup, "The Paleobiological Revolution"

---

ing a number of the underlying assumptions of the Neo-Darwinian perspective.[1]

Margulis was intrigued by the fact that the mitochondria in cells have their own genes. Mitochondria are the cells' intracellular power factories and supply the energy for all living metabolism. Standard theory had it that only the genes in the nucleus of cells had any importance; the genes in mitochondria were considered irrelevant. Margulis eventually realized that mitochondria were once free-living bacteria that had been incorporated into cells to power their metabolism. This, and other discoveries, led to her revolutionary understanding of the nature of the evolution of complex life-forms.[2]

Margulis discovered that all complex life developed from an original symbiosis of four different bacteria: archaebacteria, spirochetes, cyanobacteria, and oxygen-breathing bacteria. After this early unification, other kinds of bacteria were incorporated into the structure of cells. Genetic mapping and comparison to free-roving bacteria have proved that three of these bacterial forms were incorporated into the

---

Any organism, if not itself a live bacterium, is then a descendant—one way or another—of a bacterium or, more likely, mergers of several kinds of bacteria. Bacteria initially populated the planet and have never relinquished their hold.

—Lynn Margulis and Dorion Sagan, WHAT IS LIFE?

first nucleated cells. The remaining step is proving that spirochetes, or wriggling bacteria, were incorporated into cells to give them mobility. In essence, all nucleated cells were formed from the fusion of individual bacteria. Unlike individuals joined together to form entirely different, more complex entities. These new organisms had the original characteristics of the simpler bacteria as well as unique qualities that came from the synergy of their fusion. This process of evolutionary fusion went on over time to produce increasingly more complex forms of life. Margulis found that evolutionary novelty arises from symbiosis or mutual collaboration between differing life-forms; that evolution is the emergence of individuality from the interblending of once independent organisms.

She also realized that all taxonomic classifications were incorrect and she corrected this error with her "Five Realm Taxonomy of Life," basically acknowledging that bacteria are the foundation of all life on Earth.

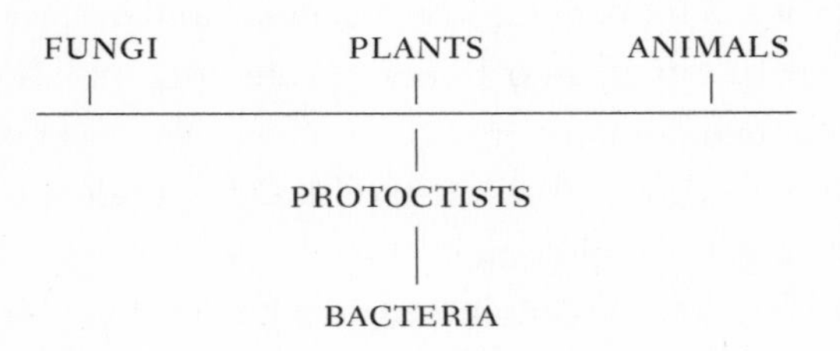

Bacteria are, in essence, the primary life-form on Earth. They combined to form nucleated cells that then fused into more complex forms: plants, fungi, animals. Many of the free-living bacteria that blended together during symbiosis can still be found within us, within all animals and plants. Chloroplasts—originally cyanobacteria—were incorporated into the majority of plant cells on Earth. It is the chloroplasts that make plants green, engage in photosynthesis, and turn power from the sun into the food and energy that enables plants to live. The neurons and nerve cells, axons and dendrites in our brains contain the same microtubules that make up the bodies of spirochetes, or wriggling bacteria. As Margulis notes: "The oxygen we breathe enters the brain from our bloodstream and is incessantly metabolized by the mitochondria that we know are former respiring bacteria. . . . We remain symbiotic beings on a symbiotic planet."[3]

Human beings are a complex symbiosis of bacteria, as are most other life-forms on Earth. We are in fact, as many indigenous people have insisted, relations with all other life-forms. The innate feeling for life that is basic in human beings, biophilia, comes both from our similarity of nature and our long close association throughout the emergence of our species. Or as Edward O. Wilson says:

> *Other species are our kin.* This statement is literally true in evolutionary time. All higher eukaryotic organisms, from flowering plants to insects and humanity itself, are thought to have descended from a single ancestral population that lived about 1.8 billion years ago. Single-celled eukaryotes and bacteria are linked by still more remote ancestors. All this distant kinship is stamped by a common genetic code and elementary features of cell structure. Humanity did not soft-land into the teeming biosphere like an alien from another planet. We arose from other organisms already here.[4]

Wilson's description of the nature of biophilia recognizes that it is primarily an *emotional* affiliation with other *life,* not a mental process of recognizing the connections between bits of the mechanical parts of the Universe that happen to inhabit a ball of rock in space. We are, by species history and genetic tendency, encoded for recognition of the aliveness of the world and an emotional bonding with it.

The Importance of Bonding

Bonding—the uniting psychological force in families—begins with our mother immediately after birth. We are placed on her stomach, our mouth near the nipple, exhausted from our long journey. She looks down and our eyes make contact. We begin to nurse. Once surrounded by the fluids of her body and the skin of her womb, we are now surrounded by her smell, her touch, the light reflected off her body; we are filled with the taste of her milk.

This looking, touching, smelling during the initial days and weeks of life initiates a deeply uniting psychological force between our mothers and ourselves. We know we are of the same family, the same blood. And the bond naturally extends outward with time. It expands to our father; we come to know his smell, his touch, the sound of his breathing, the feeling as he enters a room. We bond with other family members: brothers and sisters, our grandparents, aunts and uncles, cousins, and great-grandparents. We come to know their touch, their smell, the sounds of their voices. And as we grow, our capacity for bonding expands still further to family friends, to our neighbors and neighborhood, to our culture. Historically, this expansion naturally continued outward, beyond people, to the land upon which we live, to Earth itself.

Fathering and mothering [are] afforded by the world every day in what it sends our way. . . . Children, especially, recognize this nurturance and instruction offered by nature. According to the observations of the brilliant pioneer of ecology Edith Cobb, the imagination of children depends wholly on this contact with the environment.

—James Hillman, THE SOUL'S CODE

One of the first movements toward bonding with Earth is when we learn to crawl. It happens early, perhaps at nine months of age. We have gained some control over our bodies and have some mobility. For the first time, of our own volition, we begin to leave our mother's side. We begin to crawl, across the floor or outside on the grass, until some internal recognition tells us that we have gone too far. Frightened, we return to our mother, needing to know she is still there. She holds us, tells us we are good and that we are loved, that everything is all right. Pretty soon we want to leave again, to explore this world we are in.

Our eyes move over the landscape; it is filled with interesting colors. Our ears listen; Earth is rich with sounds. We begin to put things in our mouths, tasting the Earth, sticks, and grass. We smell the scents in which we swim, the smells of plants, and dogs and cats. Our hands feel the landscapes that we travel across. We feel the emotions that these things generate. All of this finds a place inside us; we store it away. Biophilia begins to awaken.

As children age the process extends itself. The four-year-old may spend a lot of time playing alone in the yard, sitting beneath a tree, or talking to flowers. By eight years of age the first period of lengthy time in nature without adult supervision has occurred. Perhaps it is an hour spent catching minnows in a creek, or a morning chasing frogs, or a day

---

If you kill off the prairie dogs there will be no one to cry for rain.

—Navajo warning

Amused scientists, knowing that there was no conceivable relationship between prairie dogs and rain, recommended the extermination of all burrowing animals in some desert areas planted to rangelands in the 1950s "in order to protect the sparse desert grasses. Today the area (not far from Chilchinbito, Arizona) has become a virtual wasteland."

—Bill Mollison, PERMACULTURE

Water under the ground has much to do with rain clouds. If you take the water from under the ground, the land will dry up.

—Hopi elder

Burrowing creatures, such as prairie dogs, open millions upon millions of tubes in the soils of Earth. As Mollison notes, these "burrows of spiders, gophers, and worms are to the soil what the alveoli of our lungs are to our body." As the moon passes overhead the underground aquifers rise and fall and Earth breathes out moisture-laden air. This exhalation of negative-ion-charged air through the many fissures and tubes opened by the burrowing creatures helps create rain.

How could indigenous peoples have known this? By all our standards of scientific knowledge they could not. We have neglected to realize that indigenous peoples have always had access to the finest probe ever conceived, one that makes scientific instruments coarse in comparison, one that all human beings in all places and times have had access to: the focused power of human consciousness.

collecting interesting feathers and rocks. The necessity for unsupervised time in wild landscapes is important during this later time. The accumulating years of exploratory contact and experience of nature coalesce into biophilia, into a deep bonding with Earth. The regular contact with nature, growing longer and more unsupervised as time goes by, allows the genetic predisposition for biophilia to awaken.

The continual immersion in nature where the bonding process is supported and encouraged allows it to deepen into *biognosis*—direct, depth knowledge of nature that cannot be reduced to the assembly of a collection of bits of accumulated information. Knowledge of the complex interactions of natural systems or the contribution of individual members is gained without being able to pinpoint each step in the process. There may in fact be no steps; it comes in dreams or a flash of understanding. The knowledge, because of immersion in biophilia, is directly communicated from the landscape, plants, or animals themselves. There may be a predicating factor that bursts the knowledge into awareness, but the many elements that went before are and remain unconscious—an expression of the ancient interplay between organisms interwoven in the matrix that gave them birth as species—an interplay between species that are, at their core, relations.

Unfortunately, biophilia and biognosis are being interrupted throughout the world, especially in Western cultures. The loss of biophilia leads directly to the loss of biophilia-dependent biognosis and this creates large-scale risks for us as a species. Certain holistic aspects of the workings of Earth and its ecosystems cannot be understood without utilizing the natural and ancient human capacity for biognosis—the

---

There are several ways not to face life: by taking drugs, watching television, becoming a fakir in a cave, or reading pure science. All are an abdication of personal responsibility for life on earth (including, of course, one's own life). Value- and ethic-free lifestyles are as aberrant in science as in society.

—Bill Mollison, PERMACULTURE

grasping of gestalts of whole-system functioning. Without the natural emergence of biophilia and biognosis it is unlikely that our species will be able to continue to successfully inhabit Earth. Without an activated capacity for empathy with other living things we risk becoming ever more disruptive of the ecosystem functioning of the planet.

The Loss of Biophilia

Biophilia should normally awaken in children as they grow, then continue to develop in complexity throughout the rest of their lives. Without specific initiating factors (regular early contact with an experientially living Earth, for one), biophilia does not awaken in its proper time and biognosis never develops.

(Still, it *can* awaken in later life; it is a fundamental need. Biophilia can awaken at times of intense stress or change, especially at major life shifts when prior developmental stages are being recycled, such as the movement into puberty, middle, or old age. But all too often the factors that interfered with its original emergence derail these later expressions as well.)

There are a number of factors that I think are at the root of why so few people are developing a deep bonding with the living Earth—factors contributing to the loss of biophilia. Arguably among the most powerful are: the epistemological perspective that the Universe (and Earth) is not alive but simply a machine assembled from a large number of parts; the loss of natural regular access for children to nearby wild places that contain a diversity of life-forms; public schooling; and television.

The Earth is not alive. This perspective (discussed in chapter 3), in which both adults and children in industrialized societies are embedded throughout their lives, engenders a denigration and dismissal of children's natural observations that nature is alive and filled with fellow beings. Unless blessed with relatives who understand that the perspective is incorrect, children are unlikely to receive support at home, at school, or in society for their natural perceptions and they will learn to suppress them.

But the majority of scientists, even if they give lip service to either Gaia or coevolution, still act as if the Earth were a ball of white hot, partially melted rock with just a cool crust moistened by the oceans. On the surface they see life as a thinly spread layer of organisms that have simply adapted to the material conditions of the planet. With such a view go metaphors like "the Space Ship Earth." As if humans were the crew and the passengers of a rocky ship forever traveling an inner circle around the Sun. As if the 3.6 billion years that life has existed on Earth were just a prelude to the evolution of humans and to serve as their life-support system when they chanced to come aboard. . . . This is the conventional wisdom about the Earth, and it is still taught in most schools and universities. It is almost certainly wrong and has arisen as an accidental consequence of the fragmentation of science, a fragmentation into a growing collection of independent scientific specialties. . . . The majority of practicing Earth and life scientists are specialists and even though they know that the conventional wisdom is wrong, they continue to take their views of the Earth as a planet from it.

—James Lovelock, HEALING GAIA

While much that universities teach today is new and up to date, the presupposition or premises of thought upon which all our teaching is based are ancient and, I assert, obsolete. *I refer to such notions as: a. The Cartesian dualism separating "mind" and "matter." b. The strange physicalism of the metaphors which we use to describe and explain mental phenomena—"power," "tension," "energy," "social forces," etc. c. Our anti-aesthetic assumption, borrowed from the emphasis which Bacon, Locke, and Newton long ago gave to the physical sciences, viz. that all phenomena (including the mental) can and shall be studied and* evaluated *in quantitative terms.*

—Gregory Bateson, MIND AND NATURE

The loss of access to wild places that contain a diversity of life-forms. Human beings over the course of their evolution have had frequent contact with a wide variety of wild animal and plant species and ecosystem structures. These were and are intricately bound up in what and how human beings are. Such contact generates specific types of responses in people *that have been a part of human experience since our species began.* As the number of wild animals, plants, and healthy ecosystems are depleted—as local ecosystems become homogenized—children have less and less occasion to come into contact with them. (In those instances where children could have access to wild ecosystems they often spend the majority of their time in school or watching television, and as a result have little time for interacting with the wild ecosystems around them.) There is then no activating factor to generate biophilia. Though children often do have contact with domesticated pets, house plants, and the lawns surrounding their houses, this is not the same thing. If you have ever had the opportunity to look into a wolf's or coyote's eyes, even in a zoo, it is immediately apparent that they are significantly different in nature from dogs.

The idea that animals can convey meaning, and thereby offer an attentive human being illumination, is a commonly held belief the world over. The view is disparaged and disputed only by modern cultures with an allegiance to science as the sole arbiter of truth. The price of this conceit, to my way of thinking, is enormous.

—Barry Lopez, "The Language of Animals"

A two-thousand-year-old tree or an ecosystem filled with a tumultuous, complex riot of interacting plant species *feels* markedly different from a lone sapling surrounded by the grass planted in the front yard of a new housing development, or the Norfolk pine in the corner of the kitchen. The green orderly lawns surrounding children's homes do not bear any relationship to the up-and-down, uneven landscapes filled with giant, craggy outcroppings of the immeasurably ancient stones of Earth that wild landscapes often possess.

Although often difficult for reductionists to accept, differing landscapes each project different and distinct *feeling tones.* As Masanobu

Animals are far more fundamental to our thinking than we supposed. They are not just a part of the fabric of thought: they are a part of the loom.

—Peter Steinhart, "Dreaming Elands" (in Nabhan and St. Antoine)

Fukuoka observes: "Everyone seems to believe that human thought and emotions are the products of the human mind, but I think otherwise. . . . When people see a green tree, they all think that green trees are beautiful. Trees leave a sense of peace. When the wind ripples the surface of the water, the spirit becomes restless. Go to the mountains, and a sense of the mountains arises. Travel to a lake, and one feels the spirit of the water. These emotions all arise from nature. Go somewhere where nature has been disturbed and I doubt that anything but disturbed emotions will arise."[5] Halfway around the world James Lovelock echoes these same sentiments:

> There is always at the peak of Brent Tor a sense of sacredness, as if it were a place where God and Gaia meet. The feeling is intense, like that felt in great Cathedrals, caverns, and on other mountain tops. . . . Brent Tor and places like it have a sense of peace. They seem to serve as reference points of health against which to contrast the illness of the present urban or rural scene.[6]

By not having close contact with natural ecosystems that evoke the feeling tones and meanings with which human beings have evolved, occasional, regular encounters with the wild intelligence of animals inside their own world, and regular experience of wild, untamed plants growing as they have for millions of years, something is not awakened in human beings. It is then impossible for us to understand that animals sometimes climb to the top of hills simply to look at the stars, or that when a wild herd animal is killed the rest of the herd will sometimes pause and wait until, like a sigh, the spirit of the slain animal passes.

As the loss of species and habitats spreads farther and farther it is less and less likely that children will encounter wild ecosystems with a diversity of life during the period of their development when biophilia normally awakens. Those ecosystems they do encounter are more and more likely to be unhealthy. Few people have explored what the likely result will be of children regularly encountering only diminished and unhealthy landscapes. Will any biophilia at all be generated in diminished landscapes? What kind of biophilia will it be? Will or can it be healthy?

Public schooling. Public schools are completely enveloped in the perspective that the universe is not alive. The entire curriculum, except for a rare literature course or unique teacher, contains this embedded communication. Every day for the twelve to twenty years of formal schooling children are taught that they are alone on a ball of rock hurtling round the sun and that the other residents of that ball of rock are "resources" to be used or managed.

The textbooks, their authors, and the teachers are all presented as authorities on the nature of reality, and the universe-as-machine epistemology is carefully inculcated in the children. Despite the fact that increasing numbers of scientists acknowledge that mechanistic reductionism is either incorrect or has serious limitations, it continues to be taught.

Schools are, in many respects, only employee-training centers, funded by tax dollars, for a technology-dependent society. As the society more heavily emphasizes technology there is less study of the natural world and those courses of study that do focus on the natural world are deemphasized. Colleges and universities throughout the United States illustrate this trend by the loss of professors in the natural sciences. No new generation is being trained to take their place. The problem is so severe that as an article in the *Sun* noted, there are now only three professors working, all in their seventies or eighties, who have a depth knowledge of earthworms. Conservation biologist Reed Noss comments that "many universities no longer have courses in ichthyology, herpetology, mammalogy, ornithology, taxonomy of vascular plants, bryology, entomology (except, of course, economic entomology), or other

It is worth noting that [environmental devastation] is not the work of ignorant people. Rather, it is largely the results of work by people with BAs, BSs, LLBs, MBAs, and PhDs. Elie Wiesel once made the same point, noting that the designers and perpetrators of Auschwitz, Dachau, and Buchenwald—the Holocaust—were the heirs of Kant and Goethe, widely thought to be the best educated people on earth. But their education did not serve as an adequate barrier to barbarity. What was wrong with their education? In Wiesel's words, "It emphasized theories instead of values, concepts rather than human beings, abstraction rather than consciousness, answers rather than questions, ideology and efficiency rather than conscience."

I believe that the same could be said of our education. Toward the natural world it too emphasizes theories, not values; abstraction rather than consciousness; neat answers instead of questions; and technical efficiency over conscience. It is a matter of no small consequence that the only people who have lived sustainably on the planet for any length of time could not read, or like the Amish do not make a fetish of reading.

—David Orr, EARTH IN MIND

---

courses on the identification, evolutionary relationships, and life histories of organisms."[7] Our knowledge base of the natural world, like the knowledge base of indigenous peoples, is being lost. Book knowledge is no substitute for the direct transmission of knowledge from someone who has spent thirty years interacting with nature. The result, increasing in its scope each year, is a consistent disconnection to wild nature.

Television. Television is often derided but I think, in general, for the wrong reasons. Television is of pervasive importance because it is used to fulfill a basic need that we all have: it is intricately bound up with our need to dream. Dreaming is a basic need like food, clothing, shelter, or touch. The historian of religions, Mircea Eliade, comments that

> one of the four phases of sleep is called REM (Rapid Eye Move-

ment); it is the only phase during which the sleeping person dreams. The following experiments were done: Volunteers were prevented from staying in the REM phase, but were permitted to sleep. In other words, they could sleep, but it wasn't possible for them to dream. Consequence: the following night, the persons deprived of REM tried to dream as much as possible, and if they were again prevented from doing so, they proved nervous, irritable, and melancholy during the day. And finally, when their sleep was no longer bothered, they gave themselves over to veritable "orgies of Rapid Eye Movement sleep," as if they were avid to recover everything they had lost during the preceding nights. . . . These experiments . . . confirm the organic need of man to *dream* . . . . [Man]—any man—is continually fascinated by the chronicalling of the world, that is, by what happens in his world or in his own soul. He longs to find out how life is conceived, how destiny is manifest—in a word, in what circumstances the impossible becomes possible, and what are the limits of the possible.[8]

Dreaming is necessary because human beings have an innate need to make sense of things, to understand who and what they are, to continually process and interweave the *meanings* they encounter each day into the fabric of their lives. Eliade notes that "the specific mode of existence of man implies the need of his learning what happens, and above all what *can* happen, in the world around him and in his own interior world. That it is a matter of a structure of the human condition is shown, *inter alia,* by the *existential necessity* of listening to stories and fairy tales."[9] The purpose of dreaming is to allow the unconscious mind to work with the meanings of one's life, both interior and exterior. By this process a person integrates meaning into the fabric of his character, his life takes on more and more meaning as time goes by, he deepens—becomes less shallow, more alive and real.

Most dreaming is never consciously remembered, but during all of it the unconscious works with the material offered up to it through daily living to extract meaning, to understand our place and relationship to the world, the Universe, and life. It is an organic process as integral to human health and life as food, this need to work with *meaning* through dreaming. And like all our other basic needs, it has been taken by human beings and, over time, developed into art. Before printing, dreaming was crafted as art through storytelling, eventually developing more complexity as theater. After printing, storytelling followed a new tack and developed as art through written fiction. They all involve our human capacity and need to dream.

In listening to storytelling or reading fiction, as we listen, we enter a dreaming state—a fictional dream as the writer John Gardner named it. "We read a few words at the beginning of the book or the particular story, and suddenly we find ourselves seeing not words on a page but a train moving through Russia, an old Italian crying, or a farmhouse battered by rain."[10] We enter a dream world, we forget ourselves, our conscious mind sleeps.

Each of us has had the experience of suddenly being awakened from such a fictional dream—by the insistent ringing of the telephone or a sharp knock at the door. It startles us awake as if from a deep sleep and it takes a few minutes for us to recall ourselves, for our conscious mind to begin functioning again. Yet even as we answer the phone or the door, the meanings, experiences, or people we were dreaming about linger in the mind like a taste on the tongue or the fading chords of music in the ear.

The more accomplished that storytellers and authors become, the more their stories' structures resemble the organic process of dreaming and the more deeply we enter their fictional dreams. Fiction and storytelling become *great* when, in addition, the meanings embedded within them are the deep and important meanings that we grapple with daily in our lives and nightly in the deepest core of our dreaming. Gardner touches on this when he comments: "Thus the value of great fiction, we begin to suspect, is not just that it entertains us or distracts us from our troubles, not just that it broadens our knowledge of people and places, but also that it helps us to know what we believe, reinforces those qualities that are noblest in us, leads us to feel uneasy about our faults and limitations."[11]

People have the capacity to grasp deep meaning in stories from the subtlest of cues. Gardner offers a remarkable example of a writing exercise that captures this.

> One does not simply describe a barn, then. One describes a barn as seen by someone in some particular mood, because only in that way can the barn—or the writer's experience of barns combined with whatever lies deepest in his feelings—be tricked into mumbling its secrets.
>
> [So] . . . describe a barn as seen by a man whose son has just been killed in a war. Do not mention the son, or war, or death. Do not mention the man who does the seeing. . . . [If done well the result] should be a powerful and disturbing image, a faithful description of some apparently real barn but one from which the reader gets a sense of the father's emotion; though exactly what that emotion is he may not be able to pin down.[12]

These kinds of embedded meanings build up in works of fiction as they do in dreams. The name a character possesses, the color of his shirt, the pacing and rhythm of his sentences—all change who and what he is. Simple background information imparted offhand in the story—whether a man's father were "rich, or had owned elephants"—would immediately change the nature of his character. As Gardner comments: "Subtle details change characters' lives in ways too complex for the conscious mind to grasp, though we nevertheless grasp them."[13] For it is not the conscious mind that grasps these details. It is the unconscious mind, used to working with the nature of meaning, that takes them into itself and understands their import.

Over time, storytelling branched off into written fiction, and theater branched into moving pictures. With "movies" storytelling even more closely began to resemble dreaming because dreaming contains such sounds and images. Eventually this branching dynamic produced television, but television possesses a number of significant differences compared to movies. With television, dreaming is no longer an irregular, isolated event like sleep-dreaming, the theater, or a trip to the movies (events set off in time and space as important), but a habit casually undertaken at any time one wishes. It has been flattened as well. Television's stories contain meanings that have been processed to be broad and shallow rather than specific and deep in order for them to apply to the largest number of people. And the dreaming process is continually interrupted by commercials—a unique shift in our evolutionary history of dreaming.

Dreams are by their nature individually generated and their meanings apply specifically to the person dreaming. But because we are all people some of us are very alike, and so some fictional dreams apply to many people. Books such as *The Lord of the Rings* seem so real, the dreaming so like personal dreaming, the meanings and struggles so basic to what it means to be human, that millions of people find something in it. It becomes great fiction. Such is true as well of movies, such as *Casablanca*. But people generally find these book or movie dreams on their own. Something moves them to pick up this book or read that one of all the multitudes given them by relatives or friends, to see this film rather than another. People drive to the theater or pick books up and lay them down of their own volition and that act sets them aside from daily life. They are set aside in time just as settling down to hear the tale of a storyteller is set aside in time. They occur at irregular, isolated moments. Because the dreams that storytellers embody in books usually only apply to a few people or are poorly done, the vast majority of novels rarely sell more than a few thousand copies. A novelist such as Tolkien or Hemingway, whose work moves many millions of people, is uncommon. (There is, of course, a difference between books filled with the kind of meaning that occurs in dreams and books that are essentially lengthy gossip. Gossip fills an entirely different need than that filled by dreaming. Books that are only compelling storyline are, in many respects, just another form of gossip.)

Television is to dreaming what junk food is to real food. While fat, sugar, and starch are all essential foods that the human body needs and wants, in historical time they were not present in quantity. They were integrated into the food web in fairly small amounts. Industrial technology extracts and offers them in large amounts to anyone who wants to gorge on them. And a lot of people do since they are basic bodily needs. Dreaming, like food, is a basic human need and was once present in only limited amounts. Through television, industry has made it junk food as well.[14]

Television is not a special event, set aside in time, like the theater, books, or movies. It runs twenty-four hours a day, seven days a week. There are very few people in the Western world who do not watch a lot of it, few who do not regularly watch network programming or have a satellite dish or cable, fewer still who do not have a television set at all. Television needs a lot of programming and by nature and function those programs must affect the most people possible. This can be done by consistently producing great fiction—an impossibility, there are only so many Tolkiens or Hemingways—or else by lowering the depth of the dream. Making shallow dreams that can affect the broadest number of people. Television.

As a result, people are continually exposed to dreaming that works with very shallow and homogenized meanings reflective of one particular industry and way of thinking. The deep unconscious responds to the material as it does to anything that resembles dreaming—it works with what it gets, deriving what meaning it can. Because the level of meaning is so shallow people begin to live more and more in a world with less and less access to deep meaning. More and more life is lived in front of a television dreaming dreams that have been generated only to be palatable to the most millions. Less and less does dreaming come from life itself. Children begin to take these dreams and the meanings derived from them and interweave them into the fabric of their characters. It is no wonder that teenagers feel that life is meaning-less. No wonder that, as James Hillman notes, "More children and adolescents in the United States die from suicide than from cancer, AIDS, birth defects, influenza, heart disease, and pneumonia *combined*."[15] It is no wonder that teenagers are directing the deep, unconscious rage that comes from the loss of meaning outward as well; violent television or film, contrary to the cries of politicians, has virtually nothing to do with what is happening.[16]

A further disruptive aspect of television is that the dreaming that people receive, unlike all other dreaming, is not continuous; it is interrupted by commercials—specific dreams designed to get the viewer to purchase goods and services. Commercials interrupt the process of dreaming—they awaken the dreamer every ten to fifteen minutes. This is much like letting someone go into REM sleep and then waking them up every six minutes or like sitting down to a good book and having to answer the phone six times an hour. With television it happens not once but always. This corrupts a process the unconscious expects to be continuous.

The majority of dreams on television are derived from a human-centered, universe-as-machine perspective, and as such they are also an expression of our cultural mythology of a dead universe. The only life and meaning that exist are assumed to be in human communities. And those meanings in television that are developed as representative meanings of human communities occur along a very narrow and limited range of exploration.

But regular encounter with the wild was once normal and is essential to healthy dreaming. We are historically made to need such dreaming. The homogenized shallow dreaming of television, by its nature, does not meet that basic need.

Our culture no longer recognizes that each aspect of the wild, natural world possesses its own meaning and feeling tone, a *numen*. These, encountered at random in different sequence and frequency by each human being as they mature, accumulate in number by the day, week, month, and year. They are incorporated, in an unpredictable manner, in the unconscious dreaming of each person during sleep. The unconscious weaves them into the meaning that makes up the fabric of each individual human being. Without this experience of diversity of life that has always been an integral part of human life we cannot become whole. Or, as Stephen Kellert says, "The degradation of this human dependence on nature brings the increased likelihood of a deprived and diminished existence—again, not just materially, but also in a wide variety of affective, cognitive, and evaluative respects. The biophilia notion, therefore, powerfully asserts that much of the human search for a coherent and fulfilling existence is intimately dependent upon our relationship to nature."[17]

The Separation of Children from the Aliveness of the World

The four factors I discussed above have demonstrable impacts on children's experience of the aliveness of the world. The research of Gary Paul Nabhan and Sara St. Antoine in a community of people in the Sonoran Desert along the U.S./Mexico border is illustrative. They interviewed children, parents, and grandparents of O'odham, Yaqui, Anglo, and Hispanic cultural groups. These children are not urbanized; they live near or alongside natural landscapes, albeit ones often undergoing severe environmental impacts. As Nabhan and St. Antoine note, among urbanized children the attitudes and experiences they found would be even more pronounced. Among the children they studied they found that

- 35 percent of O'odham, 60 percent of Yaqui, 61 percent of Anglo, 77 percent of Hispanic had seen more wild animals in movies or on television than in the wild.
- 58 percent of O'odham, 100 percent of Yaqui, 53 percent of Anglo, 61 percent of Hispanic reported they had never spent more than a half hour alone in a wild place.
- 35 percent of O'odham, 60 percent of Yaqui, 46 percent of Anglo, 44 percent of Hispanic reported never having collected natural objects: feathers, rocks, insects, or bones.
- 23 percent of O'odham, 40 percent of Yaqui, 38 percent of Anglo, 44 percent of Hispanic could not identify which desert plant smells strongest after a rain. (It is the common and widely encountered creosote bush.)
- 23 percent of O'odham, 20 percent of Yaqui, 15 percent of Anglo, and 16 percent of Hispanic did not know that desert birds sing more in the morning than at noon.
- 17 percent of O'odham, 20 percent of Yaqui, 0 percent of Anglo, 55 percent of Hispanic did not know that the prickly pear cactus

fruit is edible. (The fruits have been a major food source for at least 8,000 years in the region and are sold in a variety of fresh and prepared forms in the local markets.)

- 58 percent of O'odham, 100 percent of Yaqui, 46 percent of Anglo, and 44 percent of Hispanic said they learned more from books about plants and animals than from their grandparents or parents.
- 47 percent of O'odham, 100 percent of Yaqui, 61 percent of Anglo, and 55 percent of Hispanic reported that most of their learning about plants and animals was coming from school not family.
- 58 percent of O'odham, 60 percent of Yaqui, 38 percent of Anglo, and 61 percent of Hispanic children felt they had learned more about animals and plants in school than their grandparents *ever* learned.

The O'odham and Yaqui children, asked to name 17 species of local plants and animals in their native language from photographs, averaged only 4.6 names. However, their grandparents, given the same test, averaged 15.1 names. Nearly every Yaqui and O'odham at the turn of the century could have named every one. Normally, indigenous peoples are exceptionally acute observers of their natural surroundings. As an example, the Barasana Indians can name every plant in their range in every stage of growth, something that no Western-trained taxonomist can do. In addition, they note a much larger range of distinctions in their taxonomy than Western scientists—they, in fact, generally identify considerably more species.

A number of the children in Nabhan and St. Antoine's survey, when asked whether books, school, or family was producing the most information about plants and animals, replied that it was none of them. "Television," they said, was the answer.[18]

Nabhan and St. Antoine make a number of observations about the results of their survey.

- Personal, uninhibited, spontaneous interaction with nature is seldom occurring among children.
- Television and formal education run counter to and interfere with

the many-generations-successful indigenous forms of education such as apprenticeships with elders.

- The ability of children to even perceive the natural multisensory nature of landscapes is diminished by constant exposure to television and textbooks.
- Television and formal schooling have usurped direct experience of the world as a provider of information about that world, as well as usurping the experiences of storytelling and individualized personal instruction from elders.
- Nonscientific, culturally generated information about nature is routinely denigrated by the formal school system. The children are taught to relegate their elders' knowledge to a cultural wastebin; taught that it is unscientific and inaccurate.
- The children are taught only Western scientific modes of analysis and are trained that *feelings* about nature are irrelevant.

The children Nabhan and St. Antoine surveyed, like most children in the Western world, are giving more credence to Western, mechanistic perspectives than older ones, according Western scientists more value as observers and reservoirs of knowledge of the natural world than other observers, beginning to view traditional knowledge as old-fashioned and out-of-date, spending much of their time immersed in classroom and television experiences, spending less and less time in

---

A lot of what matters is the power and the feeling of the experience. . . . But when you put something in a museum, or even on TV, you can see it all right, but you're really looking only at the shell.

—Barbara Smith, Navajo educator (in Nabhan and St. Antoine)

We might learn about plants in science, how to name their parts or how they grow. But we never went the next step, which was to talk about how to care for them.

—Felipe Molina, Yaqui educator (in Nabhan and St. Antoine)

I have become concerned that formal education unintentionally encourages children to discount what can be learned at home, especially when traditional knowledge about the desert is juxtaposed with that presented in texts by authoritative science experts. Most of the desert children we talked to claimed that they had learned more about plants and animals from school than their grandparents had learned their entire lives. Why? Perhaps most of the students knew that their grandparents had not finished as many years in school as they or their teachers had, so they were obviously not as "smart." . . . Still, I was most struck by the fact that the detailed knowledge of the plants and animals held by Indian elders was not considered valuable exactly because it was not book learning! It was as though our society's high regard for zoological or botanical "facts" derived from lab experiments, books, and science films has invalidated knowledge learned by other means.

—Gary Paul Nabhan, CULTURES OF HABITAT

---

contact with wild nature, and are experiencing an escalating loss and desire for knowledge of the animal and plant world. Frighteningly, because of their classroom and television experience, many of them believe they now know *more* than their elders about the natural world even though an examination of the facts show that they do not.

The process outlined by Nabhan and St. Antoine—the denigration of older epistemologies, their loss, and the inculcation of a different epistemology in children—can be seen in any examination of historical contacts between indigenous and industrialized populations. Importantly, Nabhan and St. Antoine show that it is an ongoing process, not merely historical. The outcome for succeeding generations (and for our species as a whole) is a lessening of connection to the natural world and knowledge about its workings.

But this trend should not be surprising. If the world is viewed as merely a machine, a bunch of soulless parts, there is little reason to care for it—it is merely a resource to be used. There is then no reason for the world to be experienced as alive, for the soul of a person to leave their body and intermingle with the soul of a tree or a stone, the soul of a landscape or a plant. John Gardner remarks that this is true, too, of the fictional dream and for the same reasons: "For the writer who views his characters as helpless biological organisms, mere units in a mindless social structure, or cogs in a mechanistic universe, whatever values those characters may hold must necessarily be illusions" and hence the characters quickly lose the interest of the reader. There is no reason to care for them, no reason to enter the dream.[19]

For a variety of reasons it is common in our culture to regard *only* indigenous peoples as possessing a valuable knowledge base that is being (or has been) lost or denigrated. The truth is quite different. All cultures or subgroups, including mainstream Western cultures, who possess *any* direct experiential information about and connection to the natural world, are losing their knowledge base as universe-as-machine expands in scope. This includes the naturalists and their knowledge base that Reed Noss comments on and it also includes common or folk knowledge held within the population at large.

All human beings, not just those in indigenous cultures, have the propensity for biophilia. They all develop biognosis and a subsequent body of information about the natural world if the process is allowed to emerge organically. Because biophilia is a natural, genetic, and evolutionary process, it will continue to emerge with the same tenacity as weeds breaking through civilization's sidewalks. Its emergence will be

---

What do our students lose when we teach them how to model population viability and analyze remote sensing data, but not how to distinguish the song of the Bay-breasted Warbler from that of the Cape May, the track of the mink from that of the marten, the taste of the birch twig from that of the cherry? . . . Does it follow that they will therefore have no personal emotional ties to the land? . . . Scientific abstraction and fancy technologies are no substitutes for the wisdom that springs from knowing the world and its creatures in intimate, loving detail.

—Reed Noss, "The Naturalists Are Dying Off"

A vast pool of practical knowledge about plants, animals, and ecological patterns exists within the traditions, languages, stories, and oral histories of indigenous and peasant cultures worldwide—but this "ethno-ecological" wisdom is vanishing even more rapidly than nature itself, as rural societies face accelerating social change. Revitalizing time-tested, hands-on ways of learning about the places we inhabit may prove just as important for our future as the new ways of computers and the internet.

—John Tuxill, NATURE'S CORNUCOPIA: OUR STAKE IN PLANT DIVERSITY

---

greeted by reductionists with the same dismay that city planners greet those weeds—somewhat like a dinner host greets an uninvited, and rather unruly, guest at the table. Irrespective of cultural background, because of innate conflicts with universe-as-machine, historical carry-overs or re-emergences of biophilia and biognosis will be suppressed in any culture that embraces mechanistic reductionism as a primary cultural value.

THE SUPPRESSION OF BIOPHILIA IN MEDICINE

In the United States, one area where this conflict is playing itself out with terrible environmental results is in medicine—in the conflict between pharmaceuticals (universe-as-machine) and herbal medicines (Universe-as-alive).

European settlers in America and their descendants, like all humans everywhere, traditionally used plants as medicines, for some 350 years before the commercial availability of penicillin in 1946; they were our primary medicines. And of course, American Indians had used medicinal plants for thousands more years before that. Indeed they were, and always had been, the primary medicines of the human species. Plant medicines belong to an older biophilia-based epistemology; pharmaceuticals to a newer, scientific one. But between 1906 and 1928 scientists and physicians in the United States began systematically attacking the use of plants as medicines. In general, they asserted that the medicinal use of plants was only the remnant of an unscientific past, that this part of nature was too dangerous, wild, and unpredictable to use for healing, and that, in fact, in many instances it did not work at all and never had. They began systematically denigrating the plant knowledge of elders (indigenous or not), turning it into backwoods superstition and uneducated ignorance—the image of an old, toothless, badly trained (if at all) granny pulling out of her pocket some plant leaves or root and insisting that it will heal being contrasted with the young modern medical practitioner in his starched white coat with dangling stethoscope.

Scientific theories and perspectives were developed to support these assertions, and they have been widely, and regularly, published in scientific journals and the news media for the past century. Eventually, these theories and perspectives were widely accepted by state legislatures, the media, and, over time, the public at large. Medical schools or herbal practitioners in the United States who held to the use of plants as medicines were forced out of business by the combined power of physicians, pharmaceutical companies, and college accrediting bodies acting through legislatures, the legal system, and policing agencies. By 1936 the last botanical medical college had closed and Morris Fishbein, an early editor of the *Journal of the American Medical Association* (JAMA), would soon comment with satisfaction that "all the signs and portents indicate that the great deluge of modern scientific chemotherapy is about to wash away the plant and vegetable debris."[20]

The reductionist, universe-as-machine epistemology that replaced Universe-as-alive, offered up scientifically developed pharmaceuticals in place of plant medicines. Scientists and reductionist physicians insist they are better. (Still, in a strange parody pharmaceutical manufacturers call the building where they make their drugs a "plant.") Their reasoning is well known: Pharmaceuticals are subjected to intense scientific scrutiny, subjected to exacting double-blind studies to ensure safety, and they are more powerful and controllable in their actions than plant medicines, which often vary in their chemistry and, hence, are less reliable. (Less publicly acknowledged is that because they are man-made, pharmaceuticals can be controlled and will generate more money for corporations than plants ever will.) Pharmaceuticals and medical science together, it is asserted, can and will cure more diseases than the older epistemology of plant medicines ever could because these new ones are based on *science.*

In spite of the fact that the past twenty years has seen a resurgence of herbalism in the United States, it is still not integrated into regular medical practice, insurance companies will not reimburse for its use, and no serious article on the legitimacy of herbal medicines has to date been published by any of the mainstream news media. Scientific medicine and pharmaceuticals are accepted as the legitimate, safe, and sensible way to approach healing. Herbs are not.

Medical practitioners and scientists continue to insist that if the universe-as-machine epistemology is accepted and healing is based on the synthetic substances created through science (with the corresponding abandonment of plants as medicines), then human beings will be weller than they ever have been before. All human disease will be cured—it is only a matter of time. Or, as Dr. David Moreau wrote in 1976 in *Vogue* magazine: "If the competitive drug industry is allowed to continue the extraordinary achievements of the last sixty years, by 2036 nearly all the health obstacles to survival into extreme old age will have been overcome."[21] Considerable evidence is accumulating that they are not correct in their conclusions and that pharmaceuticals and technological medicine present the same kinds of long-term dangers to the environment that much of our technology has been found to possess.

We have had over half a century of complete reliance on this model of medicine. Let's take a look at some of the environmental and human consequences.

The Environmental Impacts of Technological Medicine | 5

What is the collateral damage
of the pharmacist's pipette?

—Dale Pendell, *Pharmako/poeia*

Industrial toxins and wastes are in the air and oceans, and buried and heaped up on the land. Our very food now seems suspect. "Better living through chemistry" was brought us by our own entrepreneurial and selectively regulatory societies. The developed world now has to work all these questions out, with vast environmental destruction as the stakes.

—Gary Snyder, "Introduction" to *Pharmako/poeia*

Average number of Zimbabweans killed each week by AIDS: 2,500

Average number of Americans killed each week by prescription drugs: 1,900

—Harper's

In their drive to conquer disease, the supporters of technological medicine have created a great many industrial products: pharmaceuticals; personal care products (things such as sunscreens and antibiotic soaps); radiopharmaceuticals and chemotherapy; and pharmaceutical delivery and medical practice products (things such as hypodermic needles, latex gloves, thermometers). All of them end up in the environment. All of them have significant impacts.

Pharmaceutical Drugs

The vast majority of pharmaceutical drugs do not heal diseases—they control symptoms by introducing chemical mediators, at specific levels, into the body. People with high blood pressure, for example, are not cured when they take medication, which is why they have to take it regularly, often for the rest of their lives. Unlike plants, blood pressure medications, and nearly all pharmaceuticals, are not a normal part of the diet nor a food previously encountered in our evolution. So, the human body excretes them throughout the day in urine and feces: 50–95 percent of each drug taken is excreted chemically unchanged or unmetabolized.[1] As blood pressure medication is excreted blood pressure begins to rise and more of the drug must be taken. Drugs used for acute conditions, such as antibiotics, are usually taken short term; those used for chronic conditions such as high blood pressure are usually taken for years or an entire lifetime. In consequence, enormous quantities of pharmaceuticals are going through people's bodies into the environment, where they are proving to have powerfully negative impacts in ecosystems. And the quantity of drugs and other biologically active medical products that are flowing into the environment is increasing every day.

A recent *New York Times* article observed that "prescription drugs are now the fastest-growing part of the nation's health care bill. That is not so much because manufacturers are raising prices for existing drugs, but because patients are switching to newly approved medicines that cost more, and more prescriptions are being written than ever before."[2] Retail prescription sales for pharmaceuticals were $42.7 billion in 1991. In 1999, a mere eight years later, sales were $111.3 billion.[3] In the next decade, as the knowledge from the unraveling of the human genome makes even more drugs possible, this figure is expected to increase substantially. At present there are some 500 known chemical receptor sites in the human body affected by drugs. With information from the human genome project this number is expected to soar to between 3,000 and 10,000 sites. As Dr. Gillian Woolett of the Pharmaceutical Research and Manufacturers Association excitedly proclaimed, "The rate of change is absolutely incredible."[4] The two scientists who have done the most research on pharmaceuticals in the environment, Christian Daughton (of the U.S. Environmental Protection Agency—EPA) and Thomas Ternes (of the Institute of Water Research and Water Technology in Weisbaden, Germany), comment that "[this] escalating introduction to the marketplace of new pharmaceuticals is adding exponentially to the already large array of chemical classes, each with distinct modes of biochemical action, many of which are poorly understood."[5]

Many excreted pharmaceuticals and their metabolites are not biodegradable and go on producing chemical effects forever. Most that do biodegrade are regularly replenished by the need for continual dosing or by new prescriptions for new people. As pharmaceuticals are excreted in pure and metabolized forms they also intermix in the waste streams that flow into the environment in ways that cannot be pre-

---

It turns out that the same qualities that make synthetic chemicals so valuable—their potency and durability, for instance—also make them dangerous. Many of them stay in their original forms for months, years, or even centuries. . . . We are all, in effect, involuntary participants in a vast, uncontrolled experiment. We have enjoyed the benefits of man-made chemicals, but we have unknowingly taken a great risk. We have unleashed tens of thousands of man-made chemicals into the air, water, and soil without knowing what the effects will be on the world's living creatures, including Homo sapiens.

—Dan Fagin and Marianne Lavelle, TOXIC DECEPTION

dicted, with effects that are not understood. Researchers have found that metabolites, chemicals produced as by-products of pharmaceutical interaction with the body, tend to be more persistent in the environment, and are sometimes more powerful in their actions, than the drugs from which they are derived.[6]

In 1999 Americans filled 2.8 billion prescriptions covering roughly sixty-six classes of pharmaceuticals. These include: antidepressants, tranquilizers, and psychiatric drugs; cancer (chemotherapy) drugs; pain killers; anti-inflammatories; antihypertensives; antiseptics; fungicides; anti-epileptics; bronchodilators; lipid regulators (e.g., high-cholesterol medication); muscle relaxants; oral contraceptives; anorectics (diet medication); synthetic hormones; and antibiotics.[7]

These pharmaceutical drugs and the personal care products also manufactured by many pharmaceutical companies (such as sunscreen lotions, lipsticks, deodorants, perfumes, and shampoos) are produced in staggeringly huge quantities; often equaling or surpassing agro-

---

Nothing conveys the amount of pharmaceuticals Americans use better than a visit to a large-chain, cut-rate pharmacy. Under flickering and humming fluorescent lights, amidst bleak metal shelving jam-packed with over-the-counter medications, a long line of people stand waiting on industrial carpet the tensile strength of Elsa Lancaster's hair in Bride of Frankenstein. Behind the counter in front of them a group of perky, white-jacketed pharmacy aides with rather doll-like expressions ring up purchases. Behind the aides, at a long table, stand four exhausted-looking pharmacists filling prescriptions. Every few minutes they leave the table and walk between shelves bowed by the weight of scores of large medication dispensers. They dial a large wheel on the front, hold a small container underneath, and the prescription pills tumble out. And though I live in a small town of 25,000 people we have 33 pharmacies; about one for every 800 people. This goes on in every town and city in America one way or another each and every day with greater and greater frequency each year.

chemicals in tonnage. The number of pharmaceuticals Americans consume is simply astounding. All of these go into the ecosystem, most of them through excretion into waste treatment systems.

Waste Treatment

The average person produces about 1,300 pounds of excrement (liquid and solid) per year and all of it's got to go somewhere. Until the development of cities, most human waste was routinely deposited willy-nilly in the ecosystem—much the way it is done by all other living beings on Earth—where it was then naturally recycled back into the soil. Rarely, if ever, was it excreted in any quantity into water systems. (Wild water was, as a result, historically quite different than it is now.) As people began to concentrate in larger numbers, some cultures utilized these natural, ancient patterns in handling their waste. Many Asian nations have, for centuries, placed varying amounts of their human waste back into the ecosystem as fertilizer. But in Christian Europe of the Middle Ages a vastly different process evolved over time.

Urinating and defecating on the ground near dwellings evolved into open pit privies or outhouses, which evolved into open pit cesspools (often for larger habitations or small towns) that tended to overflow with rains, which evolved into storm sewers where excreta was funneled into—usually open—gutters where rains would drain it into rivers.

Some smaller towns or businesses in America actually built public outhouses overhanging local rivers so that excrement could drop directly into the flowing water. Many monasteries in Europe incorporated the same techniques during the Middle Ages. People in homes that did not use outhouses or have overhanging privies collected excreta in "chamber pots" that were emptied daily, usually into cesspools or open sewers. These kinds of disposal of raw sewage led to regular epidemics throughout Europe; water supplies were badly contaminated. A fear of water, of bathing or washing of any sort, soon developed.

Because of the epidemics caused by accumulating wastes, cities, during the period they developed piped water, also developed closed and buried drain systems that funneled all waste and water from homes into central collection systems. Most of these simply discharged into nearby streams and rivers. By 1862 some 136 cities in the United States were doing this, and by 1880 there were nearly 600. Cholera, long a problem from open waste collection, abated, but cities downstream of discharges suddenly found themselves struggling with typhoid. In response cities began filtering (and eventually chlorinating) the water before people drank it. Raw sewage was still not treated.

The enormous increases in population, expansion of cities, and the spreading of industrialization and technology during the nineteenth and first half of the twentieth century produced huge quantities of waste. Industrial wastes were blended with human wastes into one enormous waste stream; little of it was treated to reduce toxicity. By the 1950s American waters receiving sewage effluent had become so polluted that, in one celebrated case, a river in the Northeast caught on fire. In response, treatment plants to reduce environmental impacts began to appear.

The earliest treatment facilities, in what became known as primary treatment, simply strained out "floatables." Secondary treatment, a later innovation, speeded up the decomposition of wastes by oxygenating them and promoting bacterial growth. The water, cleaned of everything treatment facilities could clean it of, flowed back into ground waters and streams. Left behind was a dewatered, sticky, mudlike, black goo called sludge.[8]

By 1997 industry was dumping an estimated 240 million pounds of wastes with "hazardous components" alone into municipal treatment systems and American households were contributing a staggering 1.6 trillion gallons of waste-filled water to the treatment stream. And in spite of the expansions in waste treatment in the past five decades many pollutants (including significant amounts of excreted pharmaceuticals) remain in the water. Some two dozen major U.S. utilities release so much effluent to local waters that their discharges sometimes equal half the receiving streams' volume, basically only diluting discharged effluent 2:1. And much of the effluent, in spite of treatment, is still polluted. In 1999 the Congressional Research Service commented that

> states report that municipal discharges are the second leading source of water quality impairment in all of the nation's waters (rivers and streams, lakes, and estuaries and coastal waters). Pollutants associated with municipal discharges include nutrients . . . , bacteria and other pathogens, as well as metals and toxic chemicals from industrial and commercial activities and households.[9]

Sludge contains varying amounts of any of the 70,000 different chemicals produced by industry each year as well as (according to the U.S. EPA) "volatiles, organic solids, nutrients, disease-causing pathogenic organisms (e.g., bacteria, viruses, etc.), heavy metals and inorganic ions, and toxic organic chemicals from industrial wastes, household chemicals, and pesticides."[10]

To get rid of sludge, coastal cities, the largest city-waste producers, initially dumped it in the ocean (often creating "dead zones" where nothing could live). Some inland cities put it in landfills or shipped it to the ocean for dumping. Still others incinerated it and put the resulting ash in landfills, spread it on the ground, shipped it via barge for ocean dumping, or sent it on huge container ships to Third World countries.[11] When Congress outlawed the ocean dumping of sludge in 1988 cities were faced with a huge problem: What to do with the 11.6 *billion* pounds of sludge they were producing per year. Most of them began using it as fertilizer, plowing it back into the soil—an ironic return to Asian approaches but with a vastly more contaminated product.[12]

Among other things sludge contains significant amounts of pharmaceuticals. Most of this comes from pharmaceutical companies and hospitals who direct their waste streams into municipal treatment systems, households flushing unused pharmaceuticals down the drain, people excreting pharmaceuticals they are taking, and personal care products such as sunscreens and lotions that wash off during bathing. As well, large numbers of pharmaceutical chemicals enter the waste stream from illegal drug labs, expired drugs thrown into landfills (by both households and pharmaceutical manufacturers), hospital waste (incinerated and solid), and waste produced by pharmaceutical companies during the manufacturing process.

Pharmaceutical Manufacturer Pollution

Global figures on the production of waste by pharmaceutical companies are difficult to obtain; however, Greenpeace has accumulated some sample numbers from production facilities in India. (Many pharmaceutical plants are located in Third World countries because of lower production costs and/or for environmental reasons). Three sites in India—Ankleshwar, Nandesari, and Vapi—contain a total of 5,300 industrial units (factories). Ankleshwar, which is typical, produces 250–270 million liters of liquid waste *per day* and 100 million pounds of solid waste per year. Fifty-eight percent of the waste comes from dyes and dye manufacture, 19 percent from drugs and pharmaceuticals, 5 percent from organic chemicals, and 8 percent from various products such as pesticides. Production facilities that make these kinds of products tend to be concentrated in one location because the phenols and oils derived from coal tar in the first stage of the process—organic chemicals—are the starting chemicals for dyes, pesticides, and pharmaceuticals. All the waste streams are usually then blended into one for disposal. Approximately 50 million liters of liquid and 10,000 tons of solid waste are generated from pharmaceutical manufacturing at this one production location.

> *Planetary sinks, including the atmosphere and oceans, are filling up with [industrial] wastes they cannot assimilate.*
>
> —David Orr and David Ehrenfeld, "None So Blind"

Greenpeace discloses that in these Indian locations:

> The results of this investigation and previous work by Greenpeace clearly show that Common Effluent Treatment Plants (CETPs) fail to deal with all the chemical pollutants produced by industry. By combining wastewaters from a large number of industrial units, often engaged in very different manufacturing processes, highly complex effluents and wastes, with wholly unpredictable toxicological properties, may be

generated. At best, CETPs serve to concentrate persistent pollutants from liquid waste streams into highly contaminated sludges. Our results also demonstrate that not all the heavy metals and other persistent organic pollutants are removed from the effluent resulting in significant, direct discharges to waterways outside the plant.[13]

Ground water and local waterways surrounding the plants were found to contain complex combinations of chemicals such as trichloroethene, benzene, chlorobenzene, and 1,3- and 1,4-dichlorobenzenes. Dichlorobenzene, for example, is a persistent chemical resistant to microbial degradation.

Greenpeace found similar problems with a manufacturer in England that produces a range of phenols and oils used for pharmaceutical production. Greenpeace comments that "in October 1991 the Ministry of Agriculture, Fisheries and Food found some of the highest levels of dioxin ever recorded in the UK in milk from farms surrounding the plant." Dioxins were entering the environment both from direct discharge into local rivers and through incineration at the plant.[14]

In their final manufactured form the environmental impact from pharmaceuticals continues through excretion, hospital waste streams, and landfill dumping of expired drugs. Pharmaceuticals are inserting significant quantities of highly bioactive chemicals into soils and water throughout the world.

Pharmaceuticals in the Environment

In 1992 German researchers, looking for herbicidal pollution in ground water, were surprised to find high levels of clofibric acid—a drug used to lower cholesterol levels in the blood. Subsequent studies indicated that the North Sea contained roughly 150,000 pounds of clofibric acid. The Danube River and the Po River (in Italy) were found to contain the same proportional quantities and the tap water in Berlin regularly tests between 10 and 165 parts per trillion. Many Swiss lakes and streams were also found to contain the drug. Because atmospheric transfer was systematically ruled out and Switzerland does not manufacture the drug, the environmental presence of clofibric acid is now known to come solely from prescription drug intake and excretion.[15]

As concern for pharmaceuticals in the environment increased many researchers found their work impeded by the large numbers of pharmaceuticals they were finding and the cost of identifying them. Each researcher must obtain an exceptionally expensive "library" of the chemical profiles of all pharmaceutical drugs so that when they find an unidentified chemical profile through gas chromatography they can match it to one in the library. To date, German scientists have found anywhere from thirty to sixty pharmaceuticals in water samples they have examined in numerous countries—tap, surface, and ground water.[16]

Most pharmaceuticals are designed to resist breakdown, to persist, so that they can carry out their metabolic regulatory activities without interference from the body. In consequence many are extremely long lived. Researchers have tracked one plume of contaminated ground water from a landfill at Jackson Naval Air Station in Florida that has been slowly moving underground for more than twenty years. It still contains such drugs as pentobarbital, meprobamate, and phensuximide—a barbiturate, a tranquilizer, and an anticonvulsant.[17]

Approximately one-third of pharmaceuticals are also designed to be only lipophilic—not water soluble—so that they only dissolve in fat. Lipophilic substances can readily pass through cell walls and chemically act inside them. In the environment these substances tend to concentrate in the food chain in the stored fat of all creatures. In consequence, carnivores higher up the food chain ingest increasingly large amounts of concentrated pharmaceuticals.

The top ten prescription medications in the United States (by number of prescriptions dispensed) are: Premarin (a conjugated estrogen hormone made from pregnant mares' urine), Synthroid (a synthetic thyroid hormone), Lipitor (a cholesterol-lowering drug), Prilosec (a protein pump inhibitor that stops acid secretion in treating ulcers), Hydrocodone w/APAP (a narcotic pain reliever), Albuterol (a bronchial dilator), Norvasc (for high blood pressure), Claritin (an antihistamine), Trimox (an antibiotic), and Prozac (a mood regulator).[18] Eight of these are drugs that are used for months to years at a time. And they are being prescribed in ever increasing numbers.

The sales of Lipitor rose 46 percent from 1998 to 1999 while sales of all statins (cholesterol-lowering drugs such as Lipitor) rose 20 percent. Lipitor costs $3 per pill and is taken once per day; Pfizer reported sales of the drug in 1999 at 3.56 billion dollars. Approximately 5 million Americans take statins daily and physicians and pharmaceutical companies want to raise that number, perhaps to as high as 20 million—an increase of 400 percent—to treat what they feel are the many undiagnosed people who need it. Although physicians such as Dr. Antonio Gotto, dean of Cornell University Medical College, call statins "very safe drugs"[19] for treating high cholesterol, he, like other researchers, is only looking at short-term effects in humans without any reference to their environmental impact.

Normally demonized by the press, cholesterol is actually an important natural substance, essential for all cellular functions in all animals. Statins such as Lipitor inhibit an enzyme—HMG CoA reductase—that is needed for cholesterol production to take place in living organisms. Like clofibric acid, the cholesterol-lowering drug found by Germans in water in Europe, Lipitor and other statins—taken daily for many years by millions of people—are flowing into the environment in huge quantities, where they will continue to inhibit HMG CoA reductase in whatever organisms ingest it.

What will it mean to raise our babies on water contaminated with low levels of birth control drugs and athlete's foot remedies plus Viagra, Prozac, Valium, Claritin, Amoxicillin, Prevachol, Codeine, Flonase, Ibuprofen, Dilantin, Cozaar, Pepcid, Albuterol, Naproxen, Warfarin, Ranitidine, Diazepam, Bactroban, Lotrel, Lorazapam, Tamoxifen, Mevacor, and dozens of other potent drugs, along with hair removers, mosquito repellents, sunburn creams, musks and other fragrances? No one knows, but evidently we're going to find out, learning by doing.

—Peter Montague, "Headline: Paydirt from the Human Genome"

Although most pharmaceuticals are designed to target specific metabolic pathways in humans and domestic animals, they can have numerous often unknown effects on metabolic systems of nontarget organisms, especially invertebrates. Although many nontarget organisms share certain receptors with humans, effects on nontarget organisms are usually unknown. It is important to recognize that for many drugs, their specific modes of action even in the target species are also unknown. For these drugs, it is impossible to predict what effects they might have on nontarget organisms.

—Christian Daughton and Thomas Ternes,
"Pharmaceuticals and Personal Care Products in the Environment"

All manufactured drugs end up in both water and soil. Each excreted drug is (usually) heavily diluted by these mediums, generally being found in parts per million (ppm), parts per billion (ppb), or parts per trillion (ppt). Present in measurable quantities in most rivers, streams, lakes, and commercial water supplies, they reach the highest levels in the effluent streams coming from waste treatment plants. And they do have measurable effects.

For example, Chris Metcalf, a researcher at Trent University in Ontario, Canada, detected estrone (a type of estrogen) levels in wastewater effluent up to 400 ppt and the synthetic hormone ethinylestradiol (from birth control pills) up to 14 ppt. (He found anticancer agents, psychiatric drugs, and anti-inflammatory compounds as well.) Metcalf exposed Japanese medakas (a type of fish) for 100 days to concentrations typical of wastewater streams. At concentrations of 0.1 ppt of ethinylestradiol and 10 ppt estrone the fish began to exhibit intersexual changes (showing both male and female characteristics). At 1,000 ppt all the males transformed into females.

Male fish possess a gene that, in the presence of even ppt of estrogen, will produce vitellogenin. Vitellogenin is a protein found in female fish that is responsible for making the yolk in their eggs. Male trout, caged and placed by English researchers in lakes where wastewater effluent streams discharged, showed—after only two to three weeks' exposure—increased levels of vitellogenin in their blood. Researchers near London had become alerted to the problem when they began finding male fish whose testes were loaded with eggs.[20] In Canada, Metcalf has found intersex white perch in the Great Lakes near Ontario. And American researchers have found that male walleyes living in effluent streams have quit producing sperm while male carp are showing slowed sperm motility.[21] Abnormal reproductive development has also been found in alligators and birth defects in birds—all traced to estrogen or similar endocrine disrupting chemicals.[22] Concern about the pervasiveness of reproductive- and endocrine-disrupting chemicals in the environment has become so high that on May 30, 1996, an international panel of scientists issued The Erice Report—a joint statement of alarm about the potential long-term effects of these disruptors in the environment. The report notes that "in contrast to natural hormones found in animals and plants, some of the components and by-products of many manufactured organic compounds that interfere with the endocrine system are persistent and undergo biomagnification in the food web, which makes them of greater concern as endocrine disruptors."[23]

Estrogens such as Premarin and synthetic hormones from birth control pills are one of the most commonly prescribed pharmaceuticals

---

The England that Stephen thought so good and seemed destined to inherit is done for. The growth of the population and the applications of science have destroyed her between them. There was a freshness and an out-of-door wildness in those days which the present generation cannot imagine. I am glad to have known our countryside before its roads were too dangerous to walk on and its rivers too dirty to bathe in, before its butterflies and wild flowers were decimated by arsenical spray, before Shakespeare's Avon frothed with detergents and the fish floated belly-up in the Cam.

—E. M. Forster, Introduction to The Longest Journey
(in Kaplin, Tongues in Trees)

in the industrialized world. Nine million women in the United States take some form of estrogen daily for menopause symptom relief.[24] A large number of other pollutants that act as estrogen-like compounds are also present in the environment. They come from medical waste such as phthalates and dioxins, PCBs, and some pesticides. Estrogens are also present in naturally occurring forms in the environment, usually generated by plants (phytoestrogens). There is evidence that all these estrogens and estrogen mimics can combine together and act synergistically, producing marked effects at extremely low levels—levels that cannot be predicted from a simple addition of their presence in the environment.[25]

Given this deluge of estrogens, it should be noted that physicians have been perplexed by reduced sperm counts and slower sperm motility in American men for some time. And researchers have been perplexed as well by the younger and younger ages at which American girls are entering puberty. The average age of puberty for girls one hundred years ago was eleven. Now it is less than ten in Caucasians, less than nine in African Americans. It is not unusual to find girls as young as six or seven developing breast buds or pubic hair and 1 percent of all girls now show signs of puberty by age three.[26] Obese children, whose fat may be concentrating estrogenic chemicals in greater amounts, have been found to be entering puberty younger than thin children.[27] Rising breast and uterine cancers have also been perplexing; evidence has linked those kinds of cancers (the largest category of cancers among women) to estrogens as well.

Estrogens and birth control pills are, of course, not the only pharmaceuticals accumulating in the environment. In Germany, Thomas Ternes has found up to 1 ppb of carbamazepine—an anticonvulsive drug—and 2.4 ppb of iodine-based drugs used to improve x-ray contrast—so-called Diagnostic Contrast Media. And newer drugs, such as cellular-pump (efflux) inhibitors used to fight antibiotic-resistant bacteria, are showing up in increasing amounts as well.

Cellular-pump inhibitors prevent bacteria from ejecting antibiotics from within themselves by impeding their cellular pumping mechanisms. But all life-forms use cellular pumping to rid themselves of tox-

Theo Colburn, a senior scientist with the World Wildlife Fund, has amassed research showing that many chemicals we have set loose on the environment appear to mimic estrogen in the bodies of humans and animals, disrupting the delicate balance vital to the proper functioning of their immune, nervous, and reproductive systems. Colburn's work traces the rise in sexual and other developmental defects and malfunctions in polar bears, sea gulls, and other species across the world and behavioral deficits in children born to women who were exposed to high levels of chemicals from eating contaminated fish.

—Dan Fagin and Marianne Lavelle, TOXIC DECEPTION

The presence of numerous drugs [in the environment] sharing a specific mode of action could lead to significant effects through additive exposures. It is also significant that drugs, unlike pesticides, have not been subjected to the same scrutiny regarding possible adverse environmental effects. They have therefore enjoyed several decades of unrestricted discharge to the environment, mainly via sewage treatment works. This is surprising especially since certain pharmaceuticals are designed to modulate endocrine and immune systems and cellular signal transduction and as such (as opposed to pesticides and other industrial chemicals already undergoing scrutiny as endocrine disruptors) have obvious potential as endocrine disruptors in the environment.

—Christian Daughton and Thomas Ternes, "Pharmaceuticals and Personal Care Products in the Environment"

---

ins. Anything that interferes with cellular pumping could result in any life-form being unable to remove toxins from its cells.[28] Some evidence that this is actually occurring in the environment has already been found.

In the past decade it has been discovered that many aquatic organisms, especially bottom feeders and filter feeders (for example, shrimp, flounders, oysters), possess a special excretory system called the multixenobiotic transport system (MTS). It is composed of proteins (such as

Pgp) that facilitate the removal of toxic substances from inside their cells. Because of their nature, filter feeders and bottom feeders encounter large numbers of toxins in their diet. These types of aquatic dwellers depend heavily on the MTS, otherwise toxics would build up to deadly levels in their bodies. The MTS system is nonspecific: it recognizes many pesticides, drugs, and natural toxins alike as substances that need to be sequestered and removed.[29] Drugs such as verapamil (a cardiac calcium ion influx inhibitor) directly bind to the receptor cite of Pgp thus limiting the effectiveness of the MTS system and its cellular pumping mechanisms. As a result, at lower levels, toxins have become more dangerous to many aquatic organisms. Daughton and Ternes note that "exposure to verapamil at micromolar concentrations and lower greatly increases the toxicity of a number of drugs or other xenobiotics for many aquatic organisms as the toxicant cannot be readily removed from the exposed organism."[30] Other drugs that have been shown to inhibit the MTS include reserpine (antihypertensive), trifluoroperazine (antipsychotic tranquilizer), cyclosporins (immunosuppressants), quinidine and amiodarone (antiarrythmics), anthracyclines (noncytotoxic cytoxin analogs), and progesterone (steroid).[31]

Other drugs, such as selective serotonin reuptake inhibitors (SSRIs) like Prozac, Zoloft, Luvox, and Paxil, have exceptionally strong impacts on aquatic organisms as well—even in tiny amounts of parts per billion. Serotonin is important in invertebrate and vertebrate nervous systems but it also plays key roles in the physiologic regulatory activities of many organisms. Among shellfish serotonin regulates reproductive activities (such as spawning, egg maturation, and hatching), heartbeat rhythm, feeding, biting, swimming patterns, cilia movement, and larval metamorphosis. Among crustaceans it stimulates the release of many different neurohormones that affect such things as glucose uptake, shell color, molting, egg maturation, and levels of neuroactivity.

Some commercial shellfish farmers have long added serotonin to their crops to stimulate spawning. Researchers, however, have found that Prozac and Luvox are the most potent such compounds ever produced, having *significant* effects at parts per billion levels. Extremely low doses of Prozac initiated significant spawning activity in mussels,

Let's say you have a good cream, so you add a little vitamin A. Then you add vitamin B, then you add C and D and E. Now suppose you want to add a little selenium and a little copper, and suppose you want to add an antioxidant. Now we have a cocktail. I know a product that has 67 active ingredients.

—Dr. Albert Kligman (in Gina Kolata,
"A Question of Beauty: Is It Good for You?")

---

while Luvox was even stronger, needing several magnitudes of dosage less to produce effects. SSRIs have been found to significantly affect fingernail clams, mussels, fiddler crabs, crayfish, snails, squids, and lobsters with wide-ranging effects at extremely low doses. SSRIs such as Prozac are some of the most widely dispensed drugs in the industrialized nations. And still other drugs have been found to affect crustacean reproduction as well.

Fenfluramine, a sympathomimetic amine once popularly prescribed as a diet drug (removed from the market in 1998 because of heart valve damage in patients), has also shown strong reproductive system activity in crustaceans at low doses: It triggers ovary-stimulating hormones in crayfish and gonad-stimulating hormones in male fiddler crabs. And retinoids, prescribed in large quantities for such things as acne (Accutane), cancers such as leukemia (Vesanoid), and wrinkles (Retin-A or tretinoin, an antiaging prescription and a widely prescribed drug in the U.S.), have been shown to have profound effects on amphibian embryonic systems. Constant exposure can produce deformities in the offspring of frogs and other amphibians.[32]

Personal Care Products

Personal care products (PCPs) are essentially mild, nonprescription medications and chemicals designed to enhance personal appearance, prevent or reverse the external signs of aging, or protect the body in some fashion from the environment. They enter the environment in incredibly huge numbers. In Germany alone, in 1993, the output for

I have an image of perfect health. It's a photograph of a young woman smiling, her teeth white, her hair shining, her skin glowing, her body thin and muscular—the sort of picture that appears on the cover of magazines like American Health. *It's the true me, of course, the me if I did everything right and took an antiaging pill.*

—Gina Kolata, "A Question of Beauty: Is It Good for You?"

---

bath additives; shampoos and hair tonics; skin care products; hair sprays, setting lotions, and dyes; oral hygiene products; soaps; sunscreens; and perfumes and aftershaves totaled one *billion* one hundred and eighteen million pounds.[33]

Synthetic musks (fragrances) are present in nearly all these products. They resist biodegradation and concentrate over time in muscle, fat, and breast milk of all species in the ecosystem. They have been found in fish and human breast milk in nearly every ecosystem in the world. Because they are put on skin they tend to enter the human body directly; they also enter the environment through body washing (which inserts them into waste treatment streams) and ocean and lake swimming (which deposits them directly into ground water). Because they are aromatic volatiles and are sometimes sprayed on, musks, like some medicines such as bronchial dilators (which are inhaled), enter the atmosphere directly in aerosol form and can be uptaken by the stomata of plants (or inhaled by animals) throughout the world. Carried on wind currents, they have been found even in the supposedly unpolluted air of northern Norway. During the anaerobic activity that occurs in landfills some synthetic musks undergo transformation into highly toxic forms.[34]

Sunscreen agents are also bioaccumulative, tend to enter the water systems directly from hundreds of millions of bathers each summer throughout the world, and have been found in high concentrations in fish, especially in lakes used for recreational swimming. Other than one study finding that young girls who were regularly exposed to hair-care products containing estrogen or estrogen mimics entered puberty early, virtually no research has been done on the effects of PCPs in the ecosystem.[35]

Chemotherapy

About a million people per year receive chemotherapy in the United States, about four out of five cancer patients. (Some 600,000 of them receive radiation therapy as well).[36] A regimen of chemotherapy for a single patient can entail massive quantities of drugs over a substantial period of time. As an example, one Memorial Sloan-Kettering patient's listing of chemotherapeutic drugs covered eight pages with single-spaced print, each line a record of a course of drug treatment.[37]

Chemodrugs, antineoplastics, are extremely toxic compounds and possess mutagenic, carcinogenic, teratogenic, and embryotoxic actions. That is, they cause mutations, cancer, and abnormalities in sperm, egg, and developing embryos. Oncology nursing manuals routinely warn that chemodrugs pose significant risks to health care workers, such as skin damage, liver damage, chromosomal lesions, reproductive abnormalities, and blood problems. Environmental activist Kenny Ausubel comments that

> nurses are instructed to wear heavy protective gloves when administering them since an undiluted drop on the skin can burn a hole in the flesh. They are further advised to don face masks and long-sleeved gowns, and scrupulously avoid breathing in the agents. Disposal is accomplished in shielded containers labeled Caution: Biohazard. . . . bodily excretions saturated with the drugs are a toxic waste.[38]

People on chemotherapy excrete approximately 650,000 tons per year of bodily waste (somewhat over a billion pounds), all of which goes into the sewage stream. But researchers have found that the amounts of many of the various chemotherapy drugs entering as influent from hospitals and exiting as effluent from treatment plants are nearly identical, indicating that the drugs are relatively unaffected by sewage treatment. Many of the drugs have been found active in the ecosystem where they tend to resist microbial biodegradation. Several of the palatinate chemodrugs, such as carboplatin and cisplatin, are excreted in the urine, up to 70 percent the first day. However the remainder—at least 30 percent—remains in the body and is slowly excreted over a period of years. Researchers have shown that the loading of surface waters with genotoxic compounds occurs mostly from domestic sources, primarily hospitals and people dosed with chemotherapeutic drugs who excrete them later at home.[39]

Daughton and Ternes express concern about the presence of these drugs in the ecosystem, "not just for their acute toxicity but perhaps more for their ability to effect subtle genetic changes, the cumulative impact of which over time can lead to more profound ecologic change."[40]

Radioactive Materials

Radioactive materials are also used in treating cancer patients and play significant roles in technological medicine in general. Three main elements come into play, two directly radioactive (radiopharmaceuticals and x-rays, or teletherapy), and one not (Diagnostic Contrast Media, or DCM, which are usually nonradioactive drugs designed to help x-rays produce detailed images of soft tissues).

Diagnostic Contrast Media

About 6 million pounds of the more widely used DCMs are produced each year. They remain 95 percent unmetabolized by the human body and due to their nature are extremely resistant to transformation during waste treatment processes. Those that do transform tend to become resistant metabolites.[41] Because of this they accumulate in greater numbers each year, persisting long-term in ecosystems. The five main contrast agents have routinely been found in municipal waste treatment effluent streams in concentrations ranging from nanograms to micrograms per liter. Their long-term environmental effects are unknown.

Teletherapy

X-ray machines most often use the radioactive isotope cobalt-60 to produce their images, though a few types also use cesium-137. Both have somewhat lengthy decay rates and, due to the pervasiveness of the technology in dental offices and hospitals, present long-term environmental contamination problems. Like cancer chemodrugs, radioactive isotopes possess mutagenic, carcinogenic, teratogenic, and embryotoxic actions. X-ray equipment, when no longer needed or outdated, is discarded. In many Third World countries old x-ray machines end up in landfills where, in some instances, deaths from radiation exposure have occurred. In the United States x-ray isotopes are considered to be low-level nuclear wastes and regulations mandate they be stored in special low-level nuclear waste sites.

Radiopharmaceuticals and Low-level Nuclear Waste

While nuclear power plants are the major producer of high-level radioactive waste, industry creates the most low-level nuclear waste. Low-level waste is generally defined as radioactive materials with a half-life of 30 years or less. It is usually split into four categories, depending on the level of radioactivity: A, B, C, and greater-than-C. Misleadingly, the term "low-level" includes material with a half-life of 13 hours (such as iodine-123) or 30 years (such as cesium-137). Low-level and high-level wastes are supposed to be separated and treated at different waste sites. Nevertheless, debris classified as low-level nuclear waste is often found to contain such long-lived isotopes as technetium-99 (half-life of 210,000 years) and iodine-129 (half-life of 15.8 million years).[42]

Nuclear power plants generate many more highly radioactive low-level isotopes than industry (such as the waste products cesium-137 and cobalt-60). Nevertheless, industry often acquires these products for its own uses, usually medical. So, while the *degree* of low-level radioactivity produced by nuclear power plants is greater than for other sources, the *volume* produced by industry is higher. In 1992 the United States produced 1,745,978 cubic feet of low-level radioactive waste.

Commercial nuclear power plants produced 35 percent of the total volume, academic and direct medical uses contributed 4 percent, industry produced 52 percent, and government produced 9 percent.[43] Industrial low-level waste is composed of two primary waste types: manufacturing waste generated during pure isotope production, and waste streams from the industrial application of radioisotopes. (Raw tailings left over from the mining of radioactive ore for use in weapons development, nuclear power plants, and industrial reactors are tracked in their own category.)

Although not widely recognized by the public or reported as such by the media, the primary uses of industrial radioactive materials are medical. Industry uses radioactive isotopes primarily for irradiation (sterilization) and in the manufacture of radioisotopes used as radiopharmaceuticals, as tracers, in gauges, heart pacemakers, smoke detectors, and in x-ray equipment. Industrial irradiation includes the sterilization of medical disposables and laboratory supplies (plastic syringes, surgical gloves, nonwoven drapes and gowns, petri dishes, pipettes, and test tubes), sterilization of materials used in the pharmaceutical and cosmetic industries (e.g., raw products such as talc, cocoa butter, inorganic pigments, enzymes, starch powder, laboratory animal bedding, corks, vaccines, and manufactured pharmaceuticals), and sterilization of food supplies (as yet, a small part of industry). There are approximately 20,000 commercial users of radioactive isotopes in the United States.[44]

The primary isotopes industry uses for sterilization are cobalt-60 (half-life 5.3 years—between 60 and 100 years to decay), cesium-137 (30-year half-life—about 300 years to decay), and, to a lesser extent (mostly for food and manufactured pharmaceuticals) E-beam (which uses electricity to accelerate electrons).[45]

Isotopes are produced by three processes: electromagnetic separation, particle acceleration, and nuclear reaction. All three are usually necessary in the production of medical isotopes. Electromagnetic separation is used to produce the enriched stable isotopes that are further processed in particle accelerators and nuclear reactors. Particle accelerators generate many of the short-life radionuclides that cannot be produced in reactors. However, nuclear reactors, the most serious waste producers, are essential—"reactor produced isotopes are central to clinical [medical] practice and research."[46] Technetium-99m, for instance, one of the primary radioisotopes used in medicine, is extracted from reactor-produced molybdenum-99. Though the United States does produce some of this material itself most of it is bought from foreign sources. Basically the U.S. medical industry uses the French and other nuclear reactor systems to obtain many of its necessary isotopes.[47] The waste streams and environmental pollution from these foreign reactors, the waste from initial mining to get the reactor materials, and waste from the irradiation industry are not counted in figures that look at the contribution of medicine to radioactive waste streams (often listed as 2 percent of radioactive waste). The commercial use of radioisotopes, the amount of low-level waste, and the *degree* of radioactivity of these waste streams are all increasing yearly.

Though many of the radiopharmaceuticals used in hospitals and disease-treatment protocols tend to be short-lived, many of them *must* be produced in a nuclear facility. Nuclear facilities that produce radiopharmaceuticals have the same problems all such facilities possess.[48] Besides cobalt-60 and cesium-137 other waste products of nuclear reactors are strontium-90 (30-year half-life), nickel-59 (80,000-year half-life—750,000 years to decay), and neptunium-237 (2,000,000-year half-life—2 billion years to decay).

Cesium-137, widely used for irradiation of medical products by industry, is strongly biologically active. It biomagnifies as it moves up the food chain, being concentrated by plants and then more so by plant-eating animals, even more so by carnivores. After the Chernobyl explosion cesium-137 was found in high concentrations in sheep in England and reindeer herds in Scandinavia (leading to their destruction). It has, as well, been found in high concentrations in fish in Swedish lakes. Strontium-90 behaves in a similar fashion.[49]

Approximately 23 primary radioisotopes have been used in or are produced during clinical and biomedical research and practice.[50] They are:

Isotope	Half-life
Carbon-14	5,730 years
Radium-266	1,602 years
Cesium-137	30.3 years
Cobalt-60	5.3 years
Cobalt-57	272 days
Zinc-65	244 days
Sulfur-35	87.5 days
Iridium-192	74.2 days
Cobalt-58	70.8 days
Iodine-125	60.1 days
Chromium-51	27.7 days
Phosphorus-32	14.3 days
Iodine-131	8 days
Xenon-133	5 days
Radon-222	4 days
Titanium-201	3.4 days
Gallium-67	3.3 days
Iridium-111	2.8 days
Gold-198	2.7 days
Thallium-201	73 hours
Molybdenum-99	66 hours
Iodine-123	13 hours
Technetium-99m	6 hours

Radium-266 was the first and most common isotope used in medicine, being present in all x-ray machines through the 1950s. These machines were so common (and thought to be so safe) that many shoe stores bought them for customers to view the bones of their feet while waiting to purchase shoes. Until the mid-1950s physicians regularly used radium-266 for treating an exceptionally large variety of problems, from cancer, to depression in women, to ringworm, to x-raying fetuses prior to birth (many doctors still view pregnancy as a "problem"). Once the dangers of radium-266 were finally recognized (cancer, genetic disorders, mutations in embryos, and immune suppression, among other things) its use was curtailed and eventually it was replaced with safer, shorter-lived isotopes such as cobalt-60 and cesium-137. Waste stocks from the early extensive use of radium-266 in medicine, with a decay period of 16,000 years, are pervasive. They have been linked to a number of disorders and are known to cause, like many radioisotopes, genotoxicity.[51]

Production of medical and research radioisotopes creates regular quantities of zinc-65, cobalt-60, and cesium-137. Dismantling of cyclotron shielding materials when particle accelerators are deactivated produces disquieting amounts of Europium-152 (half-life of 13.5 years) and cobalt-60. Experimental reactor laboratories are constantly looking for new isotopes for use in medicine and generate a constant stream of waste themselves. (This includes such things as the hundreds of beagle bodies and 17.5 tons of their excrement buried at Hanford nuclear reservation from a single series of medical radiation experiments at the University of California-Davis.)[52]

The most common university research radioisotope is probably iodine-125 (approximately two years to decay), which makes up half of the average university biomedical radioactive waste (of tens of cubic yards per year).[53] Zinc-65, primarily a by-product of accelerator activation products (radiopharmaceuticals) and nuclear plant cooling, is a bioaccumulative radionuclide, and even though short-lived has measurable effects in the ecosystem. The nuclear reactors at Hanford in Washington State during the 1950s and 1960s regularly released coolant water contaminated with zinc-65 into the Columbia River. Routine testing of a worker found measurable levels in his system but it had not directly come from Hanford. His contamination was determined to have come from oysters he had eaten on the Washington coast. The water from the coolant flow had traveled, in dilute form, 360 river miles west before ocean currents carried it another 30 miles north to the bay where the oysters were harvested. It arrived in concentrations so minute as to be nearly unmeasurable. The oysters, being filter feeders with an affinity for zinc, were concentrating the radioactive element in their tissues at 200,000 times the level eventually found in the seawater.[54] Many of the radiopharmaceuticals used in medicine, in spite of

The problem of radioactive waste, however, can never be "solved" in the normal fashion. Waste cannot be destroyed, nor can scientists prove that it will stay out of the biosphere if buried. Proof of a hypothesis, via the scientific method, requires demonstration. Yet with radioactive waste, such proof would require hundreds of human generations and entail extensive risks. Critics, from aboriginal people to scientists, have often noted the presumptuousness of our civilization's willingness to reach forward in time, borrowing from the future that which we can never repay. To leave a legacy that does not merely impoverish future life but may endanger it for millennia to come, constitutes an act of unprecedented irresponsibility.

—Nicholas Lenssen, NUCLEAR WASTE: THE PROBLEM THAT WON'T GO AWAY

---

their short half-lives, do end up in our waters. The long-term environmental impacts on the gene stability of living organisms in the ecosystem are unknown.

The decay rate of radionuclides (how long it takes them to quit emitting radioactivity) is found by multiplying their half-life by a factor between 10 and 20. The usual rough estimate is based on a multiple of 10. Iodine-131, for example, which has a half-life of only 8 days, has a decay time of 80 days. These radioisotopes, when used as radiopharmaceuticals, are usually present in the body long enough for them to be excreted into toilets (in the hospital and at home) and flow through treatment plants and into water supplies or be buried in landfills. Regular quantities of radioactive material from the industry that produces medical isotopes routinely finds its way into landfills as low-level nuclear waste as well. This includes contaminated materials such as tools, clothing, cardboard, and so on. The radioactive hospital waste that is generated daily (which includes some excreted radiopharmaceuticals, contaminated paper, plastic gloves, empty bottles, glassware, cloths and bed linens, aqueous and organic liquids, liquid scintillator waste, and so on) is supposed to be tightly monitored; nevertheless, incidents where the materials find their way into the normal waste streams regularly occur.[55] Hospital radioactive waste, except for the longer-lasting isotopes (such as the cobalt-60 and cesium-137 used for irradiating some blood supplies before transfusion) is usually stored on site until it decays and then is incinerated or sent to a landfill. There is accumulating evidence, however, that many radiopharmaceuticals are contaminated with minute traces of longer-lived compounds, not measurable by normal radiation counters, that will not decay for hundreds or thousands of years.[56]

Many people who are treated in hospitals for cancer or other life-threatening diseases do not recover. In the last days of their lives a significant number of the two and one-half million Americans who die each year are dosed with staggeringly large amounts of pharmaceuticals, chemotherapy drugs, and radiation. Their bodies become a significant source of pharmaceutical pollution through crematoriums and cemeteries.[57]

Medical, Infectious, and Pathological Waste

Hospitals, veterinarians, dental offices, and physicians' clinics produce steady streams of other, nonradioactive wastes. These are generally categorized as three types: nonmedically generated hospital waste, medical waste, and infectious waste (which includes pathological waste). In the past thirty years these waste streams have increased tremendously in size because the 6,000 hospitals in the United States switched to disposable products (as have most of America's physicians, dentists, veterinarians, and population).

American hospitals produce some 6 to 7 million tons of nonmedical and approximately 3.2 million tons (6.5 *billion* pounds) of medical waste each year. Veterinarians, physicians' clinics, and dental offices produce thousands or millions of tons more. Ten to fifteen percent of medical waste, according to the U.S. Environmental Protection Agency, is infectious waste. Eighty-five percent of total waste is incinerated on-site, 15 percent is transported off-site where 10 percent of that is incinerated.[58] Nonmedical hospital waste includes things such as disposable food service items from cafeterias, office waste, paint, detergents, batteries, and shipping waste such as cardboard boxes and plastic bags. Medical waste is material that comes into contact with patients, such as IV bags, gauze dressings, syringes, bedpans, and thermometers. Waste labeled "infectious" can potentially transmit infectious disease. It is usually placed in "red bags" for easier identification. Pathological wastes are surgically removed tissues and organs and make up about 2 percent of hospital waste. This is generally incinerated on-site.

Infectious and pathological wastes can include such things as:

- cultures and stocks of infectious agents
- human blood and blood products
- human pathological wastes, including those from surgery, amputations, autopsies, etc.
- contaminated animal carcasses from medical research
- wastes from patients isolated with highly communicable diseases
- all used "sharps" such as needles and scalpels[59]

Hospitals also produce hazardous wastes, including thousands of gallons of used rubbing alcohol, xylene, formalin, tons of paint thinner, and hundreds of pounds of batteries.

Most hospital waste in the past has been incinerated; in general all medical, infectious, and pathological waste still is. Until recently, in the United States, hospital incinerators were not regulated. In consequence, many, if not most, of the incinerators had no pollution control devices. Due to increasing concern on-site incineration has decreased in favor of centralized treatment centers. Still, more than 2,000 hospitals continue to run their own incineration plants.

Hospitals, to a limited extent, also use autoclaves, microwaves, and chemical treatment to deal with medical and infectious waste. Though more environmentally friendly than incineration, these methods possess their own problems. Autoclaves present a moderate risk of occupational harm and mercury contamination of nearby surface and ground waters. Microwaves disinfect but do not sterilize the waste; a number of heat-resistant bacteria are immune. In chemical treatment the waste is ground and mixed with strong chemicals to sterilize it, then sent to landfills where the chemicals enter the environment and present their own problems.[60]

The Problem of Dioxins

When waste containing organic materials and waste containing chlorine are combined and burned, or when organic chemicals that contain chlorine are manufactured, dioxins are produced. The primary source of the medical production of dioxins is polyvinyl chloride plastics (PVCs). Because they are a staple in patient care units they are the fastest growing class of synthetic chlorinated organic compounds being manufactured. It is estimated that the production, use, and disposal of medical PVCs create the largest source of dioxin pollution in the United States. Dioxins have been found to cause cancer, immune abnormalities, birth defects, fetal death, decreased fertility, male and female reproductive dysfunction, and to interfere with various hormonal or endocrine systems (such as those that produce insulin, thyroid, and steroidal hormones).[61]

Dioxin is astonishingly potent, being active in the ecosystem in minuscule traces. In 1994 the EPA's recommended maximum exposure was 0.006 picograms/kg/per day—basically one drop of dioxin in 600,000 railroad cars full of water. (A picogram is equivalent to 1 part per trillion, ppt). That level was raised in 1997 to 0.01 pg/kg/d but current research shows that people routinely ingest 300 to 600 times that amount.[62]

Medical waste incinerators are among the top industrial sources of both dioxin and mercury contamination of the environment.

Ironically, many hospitals pollute the environment with highly toxic substances that actually contribute to public health problems. Of particular concern is the longstanding overuse of incineration for the treatment of medical waste and continued use and improper disposal of hazardous chemicals.

—Health Care Without Harm, GREENING HOSPITALS

Dioxin is highly resistant to degradation in the environment and tends to accumulate and biomagnify as it moves up the food chain. It concentrates in the fat tissues and breast milk of higher food chain animals. Nursing infants, in six months of breast feeding, will, on average, consume the EPA's recommended lifetime dose of dioxin. In extremely low doses dioxins have been shown to alter glucose tolerance and lower testosterone levels. As well, it travels quickly through ecosystems. Medical waste incinerators in Florida and Texas have been shown to be a major source of dioxin contamination in the Great Lakes.[63]

Phthalates

Another class of chemicals, phthalates, are added to PVCs in order to make them flexible and soft for use in IV bags, flexible tubing, plasma collection bags, vinyl gloves, and "sharps" containers (for scalpels and needles). Phthalates have been used extensively in things such as shower curtains, teething rings for babies, and checkbook covers.[64] Phthalates are not bound to the PVC molecules and as a result have a tendency to leach out with any exposure to liquid. This leaching is particularly acute when blood or medicines are run through IV bags and tubing into patient's bodies, or when phthalate tubing is used for kidney dialysis. Studies have shown that phthalates have carcinogenic actions and that they can damage the heart, liver, testes, kidneys, interfere with sperm production, and disrupt endocrine systems.[65] A number of drugs routinely used in IV bags, such as Taxol and Cyclosporine, have been found to significantly increase the leaching of phthalates,[66] which, if received intravenously, increases in toxicity.[67]

Mercury

Hospitals, physicians, and dentists also produce exceptionally large quantities of mercury. Mercury is used in chemical solutions, thermometers, blood pressure gauges, batteries, and fluorescent lamps—all ubiquitous in hospitals and health care centers.[68] Hospitals are currently responsible for approximately one-fifth of the mercury in waste streams in the United States. Much of it is released into the atmo-

There is a quasi-scientific fable that if you can get a frog to sit quietly in a saucepan of cold water, and if you then raise the temperature of the water very slowly and smoothly so that there is no moment marked to be the moment at which the frog should jump, he will never jump. He will get boiled. Is the human species changing its own environment with slowly increasing pollution . . . in [just] such a saucepan?

—Gregory Bateson, MIND AND NATURE

---

sphere during the incineration of medical waste. Mercury is a potent neurotoxin affecting the brain, spinal cord, kidneys, and liver. "It affects the ability to feel, see, taste and move. It can cause tingling sensations in the fingers and toes, a numb sensation around the mouth and tunnel vision [eventually leading to] personality changes, stupor and coma."[69] Mercury passes through the placenta, affecting fetal development. Wildlife populations are at risk, especially those that live on, in, or near rivers and lakes. Thirty-nine states currently regulate ingesting many of their fish species because of high levels of mercury. A number of wild bird species such as loons are exhibiting symptoms of mercury poisoning.

But of all the environmental impacts of technological medicine, none have so clearly shown such frightening implications as those created by antibiotics.

The End of Antibiotics | 6

[Once] the germ theory of contagion finally caught on, it did so with a vengeance. Different types of bacteria were implicated in anthrax, gonorrhea, typhoid, and leprosy. Microbes, once amusing little anomalies, became demonized. . . . [They] became a virulent "other" to be destroyed.

—Lynn Margulis and Dorian Sagan, *What Is Life?*

Such vehement antipathy toward any corner of the living world should have given us pause. Through our related mistakes in the world of higher animals, we should have gained the evolutionary wisdom to predict the outcome.

—Marc Lappé, *When Antibiotics Fail*

[O]ur epitaph [as a species] may well read:
"They died of a peculiar strain of reductionism,
complicated by a sudden attack of elitism,
even though there were ready natural cures close at hand."

—Gary Paul Nabhan, *Cultures of Habitat*

In 1942 the entire world's supply of chemical antibiotics was 32 liters of penicillin (isolated from a mold). By 1949, 156,000 pounds a year of penicillin and a new antibiotic, streptomycin (isolated from common soil fungi), were being produced. By 1999—*in the United States alone*—this figure had grown to an incredible 50 million pounds a year of scores of antibiotics, most of them now synthetic. As researcher W.J. Powell remarks: "In 1991, there were more than 50 penicillins, 70 cephalosporins, 12 tetracyclines, 8 aminoglycosides, 1 monobactam, 3 carbapenems, 9 macrolides, 2 new streptogramins and 3 dihydrofolate reductase inhibitors" on the market.[1] Many of those antibiotics are used in hospitals and form a significant part of their effluent stream. American physicians, outside of hospitals, dispense an additional 160 million antibiotic prescriptions each year and like all pharmaceuticals they are excreted into the environment.[2] Adding to the antibiotic waste stream, pharmaceutical manufacturers discharge tons of spent mycelial and other antibiotic-related waste into the environment, much of it still containing antibiotic residues.[3] Yearly, American factory farms dispense 20 million pounds of antibiotics so food animals will survive overcrowding and fatten for market. The inevitable excrement, in the millions of gallons, is often funneled into waste lagoons where it flows relatively unchanged into local ecosystems. Open-range farm animals (as well as millions of other domesticated animals) deposit their antibiotic-laden feces directly into the environment.[4] Ninety-seven percent, for example, of the antibiotic kanamycin passes unchanged through animal gastrointestinal (GI) tracts onto the surface of the soil.[5] In short, the American continent, like much of the world, is literally awash in antibiotics. And, as physician and researcher Stuart Levy remarks, many of these antibiotics (such as kanamycin) are not easily biodegradable.

> These antibiotics can remain intact in the environment unless destroyed by high temperatures or other physical damage, such as ultraviolet light from the sun. As active antibiotics they continue to kill off susceptible bacteria with which they have contact.[6]

In an extremely short geologic time period Earth has been saturated with hundreds of millions of tons of nonbiodegradable, often biologically unique pharmaceuticals designed to kill bacteria. Many antibiotics (literally meaning "against life") do not discriminate in their activity, but kill broad groups of diverse bacteria whenever they are used. The worldwide environmental deposition, over the past fifty years, of such huge quantities of synthetic antibiotics has initiated the most pervasive impacts on Earth's bacterial underpinnings since oxygen-generating bacteria supplanted methanogens 2.5 billion years ago. It has, as Levy comments, "stimulated evolutionary changes that are unparalleled in recorded biologic history."[7] In the short run this means the emergence of unique pathogenic bacteria in human, animal, and agricultural crop populations. In the long run experts suggest the possibility of infectious disease epidemics more potent and deadly than *any* in human history.

The Limits of Antibiotics

Perhaps no technological advance has been more widely advertised and capitalized upon than the development of antibiotics. They are routinely lauded as one of the signal elements of the successful application of science and universe-as-machine; they embody the success of the scientific method over the uninformed medicine of the past. The excitement over the discovery and successful use of antibiotics in medicine was so strong that in the late 1950s and early 1960s many physicians in the industrialized nations declared that the antibiotic era had come and jointly proclaimed the end, for all time, of epidemic disease. The Australian physician, Sir F. Macfarlane Burnet, a Nobel laureate, commented in 1963 that by the end of the twentieth century humanity would see the "virtual elimination of infectious disease as a significant factor in societal life."[8] Seven years later, Surgeon General William Stewart testified to Congress that "it was time to close the book on infectious diseases."[9] Smallpox was being eradicated and polio vaccines were showing astonishing success in preventing infection in millions of people in the United States, Africa, and Europe. Tuberculosis (TB) and malaria, it was predicted, would be gone by the year 2000. With satisfaction David Moreau observed in his *Vogue* article that "the

Although there are similarities between humans and other organisms in the acquisition of chemical defenses, there are striking differences in the deployment of those defenses. Whereas most organisms use chemical defenses to minimize their own risk of being consumed, humans use chemicals in an offensive fashion, with the express purpose of killing off not only potential consumer organisms but also potential competitors for food or shelter. Throughout history, humans have even used chemicals to kill off conspecific competitors, a use of chemicals that is certainly unusual in the natural world. . . . Most organisms manufacture complex mixtures of chemicals for defense, some of which may actually be inactive as pure compounds; humans tend to prefer highly active individual components. It is perhaps the ecologically inappropriate deployment of these natural products (and their synthetic derivatives) by humans that has led to the widespread acquisition of resistance in all manner of target species and the concomitant loss of efficacy of these chemicals.

—May Berenbaum, "The Chemistry of Defense"

Dichotomies are most mischievous when they arbitrarily separate parts of a highly interrelated and complex system. . . . Nature can be portrayed as being in opposition to us, but it also includes us; we comprise one system. Perhaps the most vivid illustration of this has been provided by Gregory Bateson, in his discussions of alcoholism and schizophrenia. Traditionally, both have been treated by forming a dichotomy—the patient on one the hand and the disease (the darker side of Nature) on the other. The two are separated conceptually, and the "disease" is treated with drugs or other therapy. Not surprisingly, the results are usually terrible; either there is no progress, or the symptoms are masked or exchanged for others.

—David Ehrenfeld, The Arrogance of Humanism

chemotherapeutic revolution [had] reduced nearly all nonviral disease to the significance of a bad cold."[10]

In spite of Moreau's optimism, when his article appeared in 1976 infectious disease was already on the rise. By 1997 it had become so bad that 3,000,000 people a year in the United States were being admitted to hospitals with difficult-to-treat, antibiotic-resistant bacterial infections; another 2,000,000 (5 percent of hospital patients) were becoming infected while visiting hospitals for routine medical procedures; and 100,000 were estimated to be dying—making infectious disease, by conservative estimates, the fourth leading cause of death in the United States. Others were not so conservative. R.L. Berkelman and J.M. Hughes commented in 1993 in the *Annals of Internal Medicine* that "the stark reality is that infectious diseases are the leading cause of death worldwide and remain the leading cause of illness and death in the United States."[11] Pathologist and researcher Marc Lappé went even further and declared in his book *When Antibiotics Fail* that "the period once euphemistically called the Age of Miracle Drugs is dead."[12]

Penicillin was first used commercially in 1945; only one year later 14 percent of *Staphylococcus aureus* bacteria were resistant. By 1950, 59 percent were resistant and by 1995 this figure had risen to 95 percent. By 1999, forty-four years after the initial commercial production of antibiotics, the first Staph strain that was resistant to all known antibiotics had infected its first three people. Evolutionary biologists had insisted that evolution in bacteria (as in all species) could only result from spontaneous, usable mutations that occur with an extremely low frequency (one out of every 10 million to one out of every 10 billion mutations) each generation.[13] That bacteria could generate significant resistance to antibiotics in only thirty-five years was considered impossible. That the human species could be facing the end of antibiotics only sixty years after their introduction was ludicrous. But, in fact, bacteria are showing extremely sophisticated response capabilities to the human "war" on disease.

We have let our profligate use of antibiotics reshape the evolution of the microbial world and wrest any hope of safe management from us. . . . Resistance to antibiotics has spread to so many different, and such unanticipated types of bacteria, that the only fair appraisal is that we have succeeded in upsetting the balance of nature.

—Marc Lappé, WHEN ANTIBIOTICS FAIL

BACTERIAL RESISTANCE

Bacteria have the capacity to generate scores of unique chemical compounds. As soon as a bacterium encounters an antibiotic, it begins to generate possible responses. This takes time, usually bacterial generations. But bacteria live a lot more quickly than we do; a new generation occurs every twenty minutes in many species, some 500,000 times faster than people. And in that quickened time scale, bacteria have found a lot of solutions to antibiotics. Among them are:

- Altering the structure of the intended internal target so that antibiotics cannot affect it.
- Decreasing the amount of the drug that reaches targeted receptors by altering its rate of assimilation, thus lowering the cellular presence of the antibiotic below toxic thresholds.
- Removing the drug from their cells as fast as it enters or preventing its entry by altering membrane structure or permeability.
- Degrading or destroying the antibiotic by creating antibiotic-specific inactivation or disabling compounds.
- Creating resistant or alternate metabolic pathways that are not susceptible to antibiotics.
- Not metabolizing the antibiotic or even changing how they metabolize it, sometimes even learning to use it for food.

Once a bacterium develops a method for countering an antibiotic, it systematically begins to pass it on to other bacteria at an extremely rapid rate of speed. In response to the pressure of antibiotics, bacteria are interacting with tremendous numbers of other bacteria and the first thing they do is share resistance information, using a wide variety of mechanisms.

- Bacteria encode several different kinds of plasmids, each of which contain resistance information, and they pass these on to other bacteria. Plasmids are self-replicating, double-stranded circles of DNA that exist independent of the bacterial chromosomes. Larger ones (R-plasmids) contain resistance information for multiple antibiotics and smaller ones (r-plasmids) usually contain information for only one. These plasmids are readily exchanged among all types of bacteria, many of which were never known to communicate before the advent of antibiotics.
- Bacteria, like most life, possess transposons, unique movable segments of DNA that are normally a component of their genome structure. Transposon activity rearranges the genetic structure of organisms and hence their biological makeup. Transposons easily move between chromosomes and plasmids and can easily integrate into DNA structures. Bacteria use transposons to transfer a significant amount of resistance information and often release them in free form into the environment to be taken up later by other bacteria.
- Integrons are a unique type of DNA sequence that integrate into the genome at nonrandom sites. They have been found to be especially active in the transfer of both resistance and virulence information. They are readily shared between bacteria.
- Bacterial viruses, or bacteriophages, also help transfer resistance information between different bacteria. It is now known that instead of making only copies of themselves when they reproduce, bacteriophages take up and make copies of host chromosome segments that contain resistance information, which are then transferred to newly infected bacteria.

Bacteria can share resistance information directly, or simply extrude it from their cells, allowing it to be picked up later by roving bacteria. They often experiment, combining resistance information from multiple sources in unique ways that increase resistance, generate new resistance pathways, or even stimulate resistance to antibiotics that they have never encountered. Even bacteria in hibernating or moribund states will share whatever information on resistance they have with any bacteria that encounter them. When bacteria take up any encoded information on resistance, they weave it into their own DNA and this acquired resistance becomes a genetic trait that can be passed on to their descendants forever. Researchers have noted that the rise of resistance over the past fifty years correlates one-to-one with the production and use of antibiotics and that *no* resistant bacteria studied before the commercial introduction of antibiotics possessed the particular resistance dynamics (or determinants) they now contain.[14]

Antibiotics, it has been found, can act as bacterial pheromones, biologically based chemical motivation signals, that literally pull bacteria to them. Once in the presence of an antibiotic, the bacterial learning rate immediately increases by several orders of magnitude. Tetracycline, in even extremely low doses, stimulates by one hundredfold the transfer, mobilization, and movement of transposons and plasmids. Powell comments that "this means that in times of stress, predicated by the presence of antibiotics, the antibiotics themselves promote the exchange of plasmids, which may contain resistance genes."[15] Bacteria also intentionally inhibit the internal mechanisms for reducing mutation in their genetic structure in order to promote quicker resistance development. Nor do bacteria compete with each other for resources, as standard evolutionary theory predicted, but rather, they promiscuously cooperate in the sharing of survival information.

The recognition, long delayed by incorrect assumptions about the nature of genetic structure, is now widespread that genetic structures in all organisms are not static but fluid, sometimes within a wide range. (This is part of a growing recognition that nature may not be red in tooth and claw but much more mutualistic and interdependently connected than formerly supposed.) Barbara McClintock, who early recognized the existence of transposons, noted in her 1983 Nobel lecture that the genome "is a highly sensitive organ of the cell, that in times of stress can initiate its own restructuring and renovation."[16] She noted as well that the instructions for how genotype reassembled came not only from the organism but from the environment itself. The greater the stress the more fluid and specific the action of the genome in responding to it. This has had a great many unlooked-for consequences.

After placing a single bacterial species in a nutrient solution containing sublethal doses of a newly developed and rare antibiotic, researchers found that within a short period of time the bacteria developed resistance to that antibiotic *and* to seven other antibiotics that it had never before encountered—some of which were structurally dissimilar to the first. Stuart Levy observes that "it's almost as if bacteria strategically anticipate the confrontation of other drugs when they resist one."[17] In essence, bacteria are anticipating the creation of antibiotics that scientists haven't even thought of yet. They are also teaching themselves how to become more virulent, how to make their diseases stronger, by sharing virulence factors among themselves through the same mechanisms they use to share resistance information. In fact, they are acting in concert so well in response to the human "war on disease" that Levy remarks that "one begins to see bacteria, not as individual species, but as a vast array of interacting constituents of an integrated microbial world."[18] Former Federal Drug Administration Commissioner Donald Kennedy echoes this observation when he states that "the evidence indicates that enteric microorganisms in animals and man, their R plasmids, and human pathogens form a linked ecosystem of their own in which action at any one point can affect every other."[19]

These molecular studies of genetic change have recast genomes as "fluid" rather than fixed, and have revealed an ever-growing list of cellular biochemical systems capable of restructuring DNA molecules and thereby reorganizing the genetic information they encode. Many natural genetic engineering systems have no other function in the normal life of cells than to mediate DNA rearrangements.

—J.A. Shapiro (in Powell, "Molecular Mechanisms of Microbial Resistance")

The trend in bacterial development of antibiotic resistance is not unlike the increasing resistance of agricultural pests to pesticides. In 1938, scientists knew of just seven insect and mite species that had acquired resistance to pesticides. By 1984 that figure had climbed to 447 and included most of the world's major pests. In response to heavier pesticide use and a wider variety of pesticides, pests have evolved sophisticated mechanisms for resisting the action of chemicals designed to kill them. Pesticides also kill the pests' natural enemies, much like antibiotics kill the natural enemies of harmful bacteria in the body.

—Michael Schmidt, BEYOND ANTIBIOTICS

And wherever antibacterial use is high, bacterial congregation and rate of learning is also high. Heavy antibiotic usage in fact causes immediate bacterial congregation, rapid learning, and a subsequent cascade of resistance information throughout the microbial membrane, where it can be accessed at any time. Researcher J. Davies notes that "this gene pool [of resistance information] is readily accessible to bacteria when they are exposed to the strong selective pressures of antibiotic usage in hospitals, for veterinary and agricultural purposes, and as growth promotants in animal and poultry husbandry."[20] Wherever antibiotics and overcrowded or ill animal life meet in large numbers resistance cascades occur: hospitals, nursing homes, day care centers, homeless shelters, prisons, inner cities, animal hospitals, and factory farming operations.

The Spread of Resistant Disease

Pathogenic bacteria have taken advantage of heavy antibiotic use and the crowding of farm animals or sick or immune-weak humans to specialize in pathogenic niches.

Enterococcus, Pseudomonas, Staphylococcus, and *Klebsiella* bacteria take advantage of surgical procedures to infect surgical wounds or patients' blood in hospitals. *Haemophilus, Pseudomonas, Staphylococcus, Klebsiella,* and *Streptococcus* infect lungs, initiating often untreatable pneumonia in elderly patients in hospitals and nursing homes. *Pseudomonas* and *Klebsiella,* traveling into urinary passageways on catheters, initiate serious or intransigent urinary tract infections in hospital patients. They also gain entry into female nurses' urinary tracts through poor hygiene, where they rapidly mutate under the pressure of free antibiotics dispensed to such hospital personnel. *Haemophilus* and *Streptococcus* initiate serious ear infections (sometimes leading to meningitis) in day care centers. Tuberculosis (TB) is increasingly resistant and is spreading in inner cities, homeless shelters, and prisons.

About 2 billion people worldwide are thought to have latent TB, about one in three people. Two hundred million of those will become infectious (15 million in the U.S.) while 3 million a year will die. About 80 percent of those infected show some signs of antibiotic resistance. Two percent, or 40 million people worldwide, currently have an untreatable, resistant, strain.[21] TB is, in fact, becoming so difficult to treat that older approaches, such as surgical removal of the diseased lung, are sometimes being utilized.[22]

Gonorrhea has reemerged with a potent resistance it learned in brothels in Vietnam among prostitutes who were regularly given daily courses of antibiotics, and now causes 360,000 infections in the United States each year.[23] Malaria, spread by mosquitos and once considered only a disease of the tropics, kills one million people a year worldwide and is resistant to pharmaceuticals 80 percent of the time.[24]

Cholera, as well, has learned resistance to a number of antibiotics from improper dosing by physicians.[25] Even more telling, it has learned resistance to the primary drug used to kill it in the wild—chlorine. Chlorine, though naturally present in the ecosystem, rarely exists in pure form, generally being chemically bonded in such things as ordinary table salt (sodium chloride). Industry currently produces 50 million pounds a year of chemically pure chlorine. This is used in such manufactured products as organochlorines (e.g., the PVCs used in medicine) and directly in water supplies as an antimicrobial additive. Its expansive use has allowed cholera to develop resistance and sparked new outbreaks of the disease worldwide. In fact, the dilute quantities of chlorine present in most tap water expose the complex bacteria in the human GI tract to chlorine daily. *E. coli,* because it is exposed to such large numbers of antibiotics (and other antimicrobials such as chlorine) in the human GI tract, is one of the principal bacteria that learns resistance and passes it on. This information exchange is especially easy with other types of GI tract bacteria such as cholera.

Antimicrobial pressure has caused *E. coli,* not normally pathogenic, to also develop unexpected virulence capacities in such forms as the potentially deadly *E. coli* O157:H7. Epidemiologists now know, through genetic markers, that it was taught its virulence by *Shigella* bacteria. Researcher Marguerite Neill observes that "judicious reflection on the meaning of this finding suggests a larger significance—that *E. coli* O157:H7 is a messenger, bringing an unwelcome message that in mankind's battle to conquer infectious diseases, the opposing army is being replenished with fresh replacements."[26]

> *The lesson from both our agricultural and medical experience is remarkable for its consistency: Ignoring the evolutionary attributes of biological systems can only be done at the peril of ecological catastrophe.*
>
> —Marc Lappé, WHEN ANTIBIOTICS FAIL

Hospitals, where large numbers of pathogenic bacteria and antibiotics come into frequent contact, give bacteria the greatest opportunity to develop resistance *and* virulence. Researchers examining the effluent streams from hospitals have found them to contain exceptionally large numbers of resistant bacteria as well as large amounts of excreted antibiotics.[27] (The prodigious amounts of antibacterial soaps that are going into the water are stimulating resistance among many classes of bacteria as well.)[28] These antibiotics and resistant bacteria flow into the environment and spread everywhere. As Julie Gerberding of the Centers for Disease Control comments, "Once restricted to hospitals, where seriously ill patients are exposed to constant infusions of drugs, these [resistant bacteria] are now being found in the community."[29]

The introduction of antibiotics into animal feed (to promote growth in spite of overcrowding) and veterinary practice has produced the same kinds of bacterial evolution. As with human diseases, pathogenic animal bacteria have specialized: *E. coli* O157:H7 in beef, *Salmonella* in chicken eggs, *Campylobacter* in chickens, *Listeria* in deli meat. (And there are others, such as *Cyclospora, Cryptosporidium,* and *Yersinia.*) Like bacteria from hospitals, they spread quickly into the environment.

To trace the flow into the environment of resistant bacteria from farming operations, researcher Stuart Levy took six groups of chickens and placed them 50 to a cage. Four cages were in a barn, two just outside. Half the chickens received food containing subtherapeutic doses of oxytetracycline. The feces of all the chickens as well as the farm family living nearby and farm families in the neighborhood were examined weekly. Within 24 to 36 hours of eating the first batch of antibiotic-containing food, the feces of the dosed chickens showed *E. coli*-resistant bacteria. Soon the undosed chickens also showed *E. coli* that were resistant to tetracycline. But even more remarkable, by the end of three months, the *E. coli* of *all* the chickens were also resistant to ampicillin, streptomycin, and sulfanamides *even though they had never been fed these drugs.* Still more startling: At the end of five months, the feces of the nearby farm family (who had no contact with the chickens) contained *E. coli* resistant to tetracycline. By the sixth month their *E. coli* were also resistant to five other antibiotics. A similar but longer study in Germany found that this resistance eventually moved into the surrounding community—taking a little over two years.[30]

Salmonella, which is now genetically lodged *in* the ovaries (and hence the eggs that come from them) of many agribusiness chickens, can survive refrigeration, boiling, basting, and frying. To kill *Salmonella* bacteria the egg must be fried hard or boiled for nine minutes or longer. *Listeria* in deli meat can survive refrigeration. And *E. coli* can now live in both orange juice and apple juice—two acidic mediums that previously killed it. These food-borne bacteria are moving with greater frequency into human populations. The Centers for Disease Control currently estimates that 76 million Americans are sickened each year by resistant food bacteria; five thousand of them die.

The rate of increase of resistance and virulence is so fast that Stuart Levy observes that "some analysts warn of present-day scenarios in which infectious antibiotic-resistant bacteria devastate whole human populations."[31] Bacteria are, in fact, learning resistance to new antibiotics in only a few years instead of the decades that it took previously.

Resistance in the Ecosystem

Though resistance in the bacteria affecting people and farm animals has been the most publicized and studied, these bacteria are not confined to people or their food animals alone. They move freely in the ecosystem and among species. As Dr. Jeffery Fisher, in his book *The Plague Makers,* notes:

> The resistant bacteria that result from this reckless practice do not stay confined to the animals from which they develop. There are no "cow bacteria" or "pig bacteria" or "chicken bacteria." In terms of the microbial world, we humans along with the rest of the animal kingdom are part of one giant ecosystem. The same resistant bacteria that grow in the intestinal tract of a cow or pig can, and do, eventually end up in our bodies.[32]

This is especially true if antibiotics flow into water. This promotes the transmission of resistant traits throughout the environment because bacterial growth is high wherever water-related biofilms occur: on the surface of water, on stones in water, and in the sediment of ponds, rivers, and oceans. Antibiotics given to fish contact all of these regions, as does the antibiotic-rich effluent from factory farms and human waste treatment facilities. Resistance transfers from domestic to wild bacteria in these biofilm regions and it tends to persist in natural ecosystems.[33]

Daughton and Ternes comment that "a number of stream surveys documented the significant prevalence of native bacteria that display resistance to a wide array of antibiotics including vancomycin. Isolates from wild geese near Chicago, Illinois, are reported to be resistant to ampicillin, tetracycline, penicillin, and erythromycin."[34] Researchers have found sixteen antibiotics commonly present in ground water and surface waters that are detectable in the microgram per liter range.

Some researchers report that these antibiotic compounds are showing genotoxicity, that is, they are affecting the integrity of genetic structures in other life-forms. Daughton and Ternes comment that this is indeed cause for concern, as the bacteria never seem to forget what has been done to them:

> Indeed, the rampant, widespread (and sometimes indiscriminate) use of antibiotics, coupled with their subsequent release into the environment, is the leading proposed cause of accelerated/spreading resistance among bacterial pathogens, which is exacerbated by the fact that resistance is maintained even in the absence of continued selective pressure (an irreversible occurrence). Sufficiently high concentrations could also have acute effects on bacteria. Such exposures could easily lead to altered microbial community structures in nature and thereby affect the higher food chain.[35]

Salmon, catfish, and trout—all raised commercially—are heavily dosed with antibiotics and other drugs; these are often blended into their food. As the food gets wet, the antibiotics (and other compounds in the food) begin to leach into the water. As well, substantial amounts fall uneaten to the pond or seabeds, where the antibiotics are taken up by other aquatic organisms. Commercial salmon, unlike catfish and trout, are raised in the open sea in pens, speeding the flow of antibacterials throughout the oceans. Because of crowded conditions, the 55 million pounds of commercial U.S. salmon are frequently dosed with antibiotics for long periods of time. Stuart Levy comments that

> since they are deposited in the water, [antibiotics] can be picked up easily by other marine animals. Tetracycline is not rapidly degraded in fish. Thus, it is excreted in its active state in feces and deposited on the sea floor. Here, too, it remains relatively stable, out of direct sunlight, which can degrade it. Consequently, the ecological effect of this antibacterial agent in the sea is the same as it is in land animals: the long-term selection of

resistant and multiresistant bacteria in salmon and other marine life.[36]

Plant communities and soil are also exposed to direct antibiotic use, not just through effluent flows. To treat infections in monocropped fields, especially in trying to control fireblight in apple and pear orchards, antibiotics such as streptomycin are sometimes sprayed in heavy doses directly on crops. This kills not only bacteria on the plants but all susceptible bacteria in the soil itself, with cascading effects on soil integrity and health. While spraying allows potent doses of streptomycin to directly enter the ecosystem, other antibiotics, such as oxytetracycline, are sometimes injected directly into larger plants' trunks and roots much as they are with people. Resistant pathogenic plant bacteria have not surprisingly been found in soil and plant communities following such applications. In the case of streptomycin, the bacterial transposon developed by leaf blight during resistance acquisition has been found in seven wild bacterial species in the soil after application. All these bacteria now have resistance to the streptomycin normally produced by the soil fungi in the region. This same dynamic has also been found occurring in the soil under wheat plants.[37] The application of antibiotics or the spreading of antibiotic effluents in the environment is promoting resistance impacts in natural soil communities among wild bacteria, thus interfering with the normal balance of the soil biota. Agricultural practices such as liming fields, as well as industrial heavy-metal pollution, also have been found to increase the density of resistant pathogens in the soil through synergistic actions that are not well understood.[38] In a new approach to agricultural disease control researchers have started to insert bacterial resistance factors directly into the genetic structure of some plants (e.g., sugar beets), but these resistance factors have also been found to move into ecosystem bacteria.

The tremendous production of antibacterial substances once found only in minute quantities in the environment—substances produced by soil fungi, bacteria, or plants to protect their territorial integrity—has begun to affect the life cycle of bacteria and thousands of other

Treating the soil with chloropicrin will temporarily alleviate bacterial soft rot in Chinese cabbage and the daikon radish, but the disease breaks out again two years later and gets completely out of hand. This germicide halts the soft rot, but at the same time it also kills other bacteria that moderate the severity of the disease, leaving the field open to the soft rot bacteria. Chloropicrin also works against fusarium fungi and sclerotium fungi that attack young seedlings, but one cannot overlook the fact that these fungi kill other important pathogens. Is it really possible to restore the balance of nature by spraying an array of bactericides and fungicides like this into a soil populated with such a large variety of microbes?

—Masanobu Fukuoka, The Natural Way of Farming

---

organisms in the ecosystem, and subsequently is affecting the health of the soil and the planet itself.

Plant pollinators such as honeybees have already begun paying the price. Both wild and domestic honeybee populations have become infected with virulent, resistant strains of antibacterial-generated pathogens. This has caused large-scale deaths of honeybee populations, with heavy impacts on the wild and domesticated plants that depend on them for pollination.

The Price We Have to Pay?

Many people insist that these environmental problems are only the price we have to pay for the best health care system on Earth. In fact, the U.S. health care system is not the best on Earth. In spite of spending more on health care per capita than any other nation (and using more pharmaceuticals), Americans rank twenty-fourth in overall *healthy* life expectancy. And in terms of length alone (as of 1997), we rank eighth in women's life expectancy, fifteenth in men's, and twentieth in infant mortality rates. Dr. Christopher Murray of the World Health Organization observes that "some areas of the United States are as bad off as impoverished African nations." People in the top countries for healthy life expectancy, such as Japan, France, and Sweden, not only live longer but spend that longer life in active health, not bed- or wheelchair-ridden like many Americans.[39] The United States ranks below Iran and Romania in the percentage of low-birth-weight babies and a poverty birth in Jamaica has a better chance of a positive outcome than one in New York City.

Despite this state of affairs, pharmaceutical and medical technology use is increasing and sales are being pushed to ever higher levels. Drug companies, for example, expanded their sales staff from 35,000 in 1994 to 56,000 in 1998—one for every eleven physicians. The companies spent 5.3 billion dollars sending their salesmen into doctors' offices in the first eleven months of 1998 and another billion for special "physician" events.[40] Media coverage plays its own part. *The New England Journal of Medicine* reported June 1, 2000, that "media reports on new medications tend to exaggerate their benefits, ignore their risks, and fail to disclose their costs."[41] Other studies have seriously questioned the improper role of bias and politics in the science of pharmaceutical development and safety studies.[42] And the *New England Journal of Medicine* has noted that not all adverse reactions to drugs can be anticipated, that 51 percent of all drugs possess side effects not detected prior to approval.[43] As a result, *properly* prescribed pharmaceuticals are now the fourth-leading cause of death in the United States. The *Journal of the American Medical Association,* in a groundbreaking study, revealed that 106,000 people each year die from side effects of pharmaceuticals and another 2.2 million (as *USA Today* reported) "suffer side effects so severe that they are permanently disabled or require long hospital stays."[44] And of course, none of this research looks at things that can go wrong; physician error and pharmaceutical contamination are only two common problems that regularly occur in scientific medicine. They often occur simply because of inescapable aspects of the human condition that will always be present in whatever century people live: incorrect assumptions and limited understandings.

In 1942, 50,000 U.S. servicemen developed acute hepatitis B from a contaminated yellow fever vaccine they were given three months earlier. And it is now known that the tremendously high incidence of hepatitis C infection in Egypt came from physicians and health care workers using insufficiently sterilized needles during inoculations for a parasitic disease (schistosomiasis). One commentator on the problem noted that "the skillful doctor began injecting at 9:20 A.M. and completed 504 injections of men, women and children by 10:10 A.M." Between 1964 and 1982, 250,000 Egyptians a year received the injections.[45] And on a larger scale, the polio vaccine administered to 98 million Americans between 1955 and 1963 is now known to have been contaminated with a simian virus, SV40. It is estimated that at least 30 million Americans were infected as a result. Monkey cells, contaminated with a virus not detectable at the time, were used in the production of the vaccine. There is growing evidence that SV40 plays a role in the development of a number of diseases, including some rare cancers.[46]

Many of the early vaccines, produced as they were with what is now thought to be ancient technology (though at the time it was leading-edge), were contaminated with many different kinds of viruses. At the time, viruses were relatively new to the tools of science; it was very hard to see them. They were not then (and, in many respects, still are not) well understood. Many vaccines were produced by removing kidneys from captive wild monkeys, chopping them up, and using the tissue to culture the vaccine virus. Louis Pascal describes the process of manufacturing oral polio vaccines (OPVs) at the time.

> Unbelievable as it may seem, the oral polio vaccine consisted of nothing more than the culture fluid from the polio-inoculated monkey kidney cell cultures after passage through a filter small enough to remove bacteria but large enough to permit passage of the polio virus, and whatever other viruses may have been infecting the monkeys before they were killed. This filtered culture fluid constituted the finished vaccine fed to hundreds of millions of people around the world. There were *no* methods used to prevent those viruses already present in the monkey kidneys from contaminating the vaccine and *no* methods used to kill the viruses after contamination.[47]

Researchers did find some previously unknown viruses—about

seventy-five of them—that naturally occur in monkey kidney tissue during the development of the polio vaccine and tried to screen the kidneys they were using with the rather crude tools then at their disposal. SV40 was not one of the viruses they could identify. Jonas Salk is phlegmatic about it. "We took all the precautions that we knew of at the time. Sometimes you find out things after the fact."[48]

A great many other similar examples exist, from the x-raying of the thymus in children (assumed to be a vestigial organ) to the wide use of Thalidomide (found safe in clinical trials) to prevent morning sickness in pregnant women. Such errors of judgment are always going to be a problem. They are compounded considerably and become especially dangerous when scientists, because of hubris and improper theoretical modeling, make incorrect assumptions about the nature of living organisms.

Bacterial Partners

Bacteria are not our enemies, as scientists postulated, nor a dangerous life-form bent on sickening humankind, as so many television commercials would have us believe. They are our ancestors and we are very much alike; we both metabolize fats, vitamins, sugars, and proteins. Lynn Margulis comments succinctly that "the more balanced view of microbe as colleague and ancestor remains almost unexpressed. Our culture ignores the hard-won fact that these disease 'agents,' these 'germs,' also germinated all life. Our ancestors, the germs, were bacteria."[49] Bacteria are not germs but the germinators—and fabric—of all life on Earth. In declaring war on them we declared war on the underlying living structure of the planet—on all life-forms we can see—on ourselves.

> *Bacteria may be Earth's tiniest life forms, but they took giant steps in evolution. Bacteria even invented multicellularity. . . . Some lineages of bacteria went on to evolve into many different kinds of beings, including ourselves.*
>
> —Lynn Margulis and Dorian Sagan, What Is Life?

Clearly, the assumptions embedded in the germ theory of disease carried hidden impacts. Accepting that theory as truth has led to behaviors—industrial, social, and environmental—that are now being recognized as having serious long-term impacts.

But Pasteur's germ theory was not the only one; there were many competing schools of thought at the time. Researchers such as Max von Pettinkofer and Elie Metchnikoff insisted that it was not the bacteria that caused disease but an interruption in the normally healthy ecology of the body that allowed pathogenic bacteria to infect it. To prove their point Pettinkofer in Bavaria, Metchnikoff in Russia, and a number of others around the world ingested liquids filled with millions of cholera bacilli. Other than experiencing a mild diarrhea none became ill. Their point was that human beings live in a sea of bacteria all the time, and the human body has learned throughout its long development to deal with them. Something must be upsetting the body's normal ability to respond to such bacilli and that is what allows them to grow unimpeded. That is the source of disease.[50] New understandings are indicating that Pettinkofer and Metchnikoff were more right than they knew.

One of the few naturally sterile places on Earth is a woman's womb, and the gestation period prior to birth is the only time any human body is bacteria free. At birth, assuming it is a healthy birth, the baby is immediately placed on the mother's chest near the nipple. During the same period that the first movements toward bonding are taking place,

---

[Our bodies] are not distinct from the bodies of plants and animals, with which we are involved in the cycles of feeding and in the intricate companionships of ecological systems and of the spirit. They are not distinct from the earth, the sun and moon, and the other heavenly bodies. It is therefore absurd to approach the subject of health piecemeal with a departmentalized band of specialists. A medical doctor uninterested in nutrition, in agriculture, in the wholesomeness of mind and spirit is as absurd as a farmer who is uninterested in health. Our fragmentation of this subject cannot be our cure, because it is our disease.

—Wendell Berry, The Unsettling of America

the bacteria that are living on the mother's skin begin to colonize her baby's body. (It is perhaps no coincidence that the largest numbers of bacteria on the human body live near the nipple—in the armpit.) When the infant begins to nurse the interior of the baby's intestinal tract is colonized with bacteria as well, from the skin around the nipple and the milk itself, and these bacteria are crucially important.

Nursing introduces lactobacilli (also found in yogurt) and other bacteria such as *Bifidobacterium bifidus* into the intestinal tract of newborns, with significant effects on their health. *Lactobacillus acidophilus* bacteria create important vitamins and nutrients such as B_1, B_2, B_3, B_{12}, and folic acid in the intestinal tract. They help digest food and they also secrete natural antibiotic substances such as acidophilin, various organic acids, and peroxides that help prevent bacterial infections in the GI tract.

One to two pounds of our adult body weight is the bacteria that live on and in our bodies. The bacteria that colonize newborns have an ancient, coevolutionary relationship with human beings and are an integral part of our species' development and our body ecology. They are in fact the first line of defense of the immune system in the prevention of disease.

The skin of our bodies, and the mucosal systems of our sinus passages and intestinal tracts, are much like fresh fertile black soil to plants. Plow up such soil, disturbing the plants that grow there, and, without planting anything, the soil will soon be covered with a profusion of new growth. The bacteria that colonize our bodies are friendly, mutualistic bacteria and take up all the space on and in our bodies on which bacteria can grow. By so doing, they leave no room for other, less benign bacteria to live. (And this is why birth in a hospital is sometimes perilous; pathogenic bacteria are present in large numbers in that environment and can colonize the baby's body before the coevolutionary bacteria have a chance to—especially if the baby is removed from the mother and placed in a separate nursery.)

The bacteria on and in our bodies help us digest foods and produce nutrients that we need to be healthy. In exchange, we give them a place to live and food to eat. But the relationship goes further. *All* our coevolutionary bacteria generate antibiotic substances that kill off threatening bacteria. For example, a form of streptococcus bacteria that normally live in the human throat, over time (as they increase in number), produce large quantities of antibacterial substances specifically active against the *Streptococcus pyogenes* bacteria that cause strep throat.

Regular exposure to pathogenic bacteria teaches our bodies and our symbiotic bacteria how to respond most effectively to disease and produces higher levels of health in later life.[51] There is emerging evidence as well that human beings are *supposed* to have one or more species of intestinal worms that coevolved with us living in our GI tracts. People in developing countries who usually have these parasites rarely develop inflammatory bowel diseases. Researchers have found that the worms engage in an intricate modulation of the body's immune system that positively affects bowel health. When Americans were given the worms by a physician, a majority of those suffering inflammatory bowel disease experienced complete remission of the disease.[52] In past years in developed countries, these parasites were assumed to be evidence of disease and killed with the use of pharmaceuticals.

The truth is, we live in an ancient, healthy symbiosis with bacterial, viral, and microfaunal colonizers. Our bodies are much like the soil of Earth, covered inside and out with a broad diversity of microfauna all providing an interdependent complex of support services. When we become ill, the symbiotic relationship with the healthy bacteria and other microfauna—our body ecology—is disturbed. The underlying

---

Because parasites and hosts coevolved, the concept of a parasite-free host is an unnatural derivative of our human experience. We strive to raise animals and plants without parasites. In fact, we douse ourselves, our livestock, and our crops with tons of poisonous pesticides to maintain this artificial pest-free status. This very act demonstrates how natural the host-parasite bond really is: like the chemical bond, we have to introduce energy to break it.

—Donald Windsor, "Equal Rights for Parasites"

factor that disrupts the body ecology is the illness, not the pathogenic bacteria that take advantage of it to occupy body sites. Antibiotics do not cure disease, they simply kill off opportunistic bacteria. Without the body's ability to restore a healthy ecology, people die anyway. More than any other disease, AIDS has taught us the limitations of antibiotics and the bacterial model of disease. Irrespective of the quantities of antibiotics used, when AIDS patients' bodies can no longer reestablish their internal ecology they die. As Marc Lappé says, "it is the *body* which ultimately controls infections, not chemicals. Without underlying immunity, drugs are meaningless."[53] Ironically, as many public health historians now know, the major decreases in human mortality and disease proclaimed to be brought about by antibiotics were due more, in fact, to better public hygiene, physicians washing their hands, and better water treatment.

Because they kill off so much of the internal symbiotic microfauna along with pathogenic bacteria, antibiotics create significant changes in human microfaunal ecology and makeup. The appearance of many diseases new to humankind, such as certain nutrient deficiencies, candida overgrowth, certain chronic infections, allergies, and chronic immune suppression are now being directly linked to the distorted internal landscape that antibiotics cause.[54] Marc Lappé comments that

> lincomycin eliminates virtually all of the bacteria that require oxygen, while neomycin and kanamycin decrease the number of oxygen-requiring germs and gram-positive anaerobic ones, leading to overgrowth of *Candida albicans* and *Staphylococcus aureus*. Polymyxin can reduce native *E. coli* to the point of extinction, leaving the terrain open for staph and strep organisms. Erythromycin has a similar favorable effect on streptococci, while bacitracin and novobiocin lower both strep and clostridia. Ampicillin and clindamycin, by contrast, appear to favor the growth of *Clostridium difficile*."[55]

And it is not just humans that have coevolutionary bacterial partners but all plant, insect, and animal life. When these other life-forms encounter antibiotics their interior and exterior ecologies are disturbed as well.

Medicine's Epistemological Error

The kinds of healing that have been generated out of a universe-as-machine model are showing the same negative and long-lasting environmental impacts that are being found with other reductionistic technologies. Modern scientists and medical practitioners, by assuming that the other life-forms on Earth are not intelligent and that Earth and its life-forms can be viewed as a collection of unrelated parts, have initiated catastrophic changes throughout the living, holistic, life-form that is our planet. These changes are affecting all parts of the globe, all life-forms, and threatening the integrity of the planetary system itself. Continuing scientific efforts to create substances that bacteria *cannot* develop resistance to are, perhaps, the most dangerous actions now occurring on Earth. If bacteria had not developed resistance to antibiotics it is doubtful whether *any* life would still exist on our planet; sufficient quantities of antibiotics have been produced over the past fifty-five years to kill off all bacteria, and hence all life on Earth, many times over.

Failing to understand bacteria as our kin, the loss of biophilia in just this one area, has initiated responses from living organisms that conventional medical epistemology insisted were impossible. This profound error has not created a disease-free life with the major cause of death extreme old age, but an ecosystem in disarray and pathogenic bacteria more virulent and powerful than ever before.

For human beings to sustainably inhabit Earth it is imperative to reexamine our approaches to healing and to medicine, as we are beginning to do with the rest of our technology. Just because medicine is intended to alleviate human suffering does not mean we are exempt from the environmental consequences of using it. Medicine needs to be reexamined in light of what we now know to be true: that Earth is a living organism, not a mere collection of parts. We need to explore the development of a system of healing that is based on the capacity of human beings to have empathy for all living things, a system that comes out of biophilia, a system attached to those older epistemologies that have been abandoned, a living medicine. The use of plant medicines instead of pharmaceuticals is one element of such a living medicine.

We've only had about a hundred years of medicine based on universe-as-machine; it is not so surprising there are errors in our understanding. Earth has been producing medicines and healing disease a lot longer than people have—and Gaia is a lot better at it than we are.

"Ever since Newton and Descartes, science has explicitly offered us the vision of total control. Science has claimed the power to eventually control everything, through its understanding of natural laws. But in the twentieth century, that claim has been shattered beyond repair. First, Heisenberg's uncertainty principle set limits on what we can know about the subatomic world. Oh well, we say. None of us lives in a subatomic world. It doesn't make any practical difference as we go through our lives. Then Gödel's theorem set similar limits to mathematics, the formal language of science. Mathematicians used to think that their language had some special inherent trueness that derived from the laws of logic. Now we know that what we call 'reason' is just an arbitrary game. It's not special, in the way we thought it was.

"And now chaos theory proves that unpredictability is built into our daily lives. It is as mundane as the rainstorm we cannot predict. And so the grand vision of science, hundreds of years old—the dream of total control—has died, in our century. And with it much of the justification, the rationale for science to do what it does. And for us to listen to it. Science has always said that it may not know everything now but it will know, eventually. But now we see that isn't true. It is an idle boast."

"This is very extreme," Hammond said, shaking his head.

"We are witnessing the end of the scientific era."

"So what will happen?" Ellie said.

Malcom shrugged. "A Change."

—Michael Crichton, JURASSIC PARK

"Plants Are All Chemists" | 7

Plants are all chemists,
tirelessly assembling the molecules of the world.

—Gary Snyder, "Introduction," *Pharmako/poeia*

Plants have the ability
to produce an almost endless number
of chemical variations
on a single chemical structure.

—David Hoffmann, *Phytochemistry: Molecular Veriditas*

IN 1803 FREDERICH SETURNER ISOLATED THE FIRST INDIVIDUAL plant constituents from opium and named them *alkaloids,* some 140 million years after complex land plants created them for reasons of their own. Plant chemistry has not been studied very long in the scheme of things; it is still not very well understood.

Consider: Each of the estimated 275,000 different species of plants on Earth contains several hundred to several thousand unique chemicals. The majority of these species manifest as millions of different individuals, all of them generating different variations, sometimes significantly, on their species' chemical theme. A plant with one thousand different chemical constituents can literally combine them in millions of different ways. To compound the complexity, these combinations, added to those of other plants or of other organisms, produce synergistic results that are not predictable. Even a tiny change in dosage or combination can produce significantly different outcomes. Basically, the little that people currently know about plant chemistry is not very much. This ignorance is magnified by our tendency (because of our upbringing) to think of plants as insentient salads or building materials engaging in chemical production processes that just happened by accident and, in consequence, have no purpose or meaning. Phytoexistentialism.

Any attempt to force a wild system into the confines of a formal system is inescapably arbitrary. Wild systems and formal systems inhabit separate dimensions. And while there is a hazy, fuzzy-edged intersection of the two planes, the linkage remains metaphorical.

—Dale Pendell, PHARMAKO/POEIA

Still, here we are.

AN ARBITRARY BEGINNING

Since their emergence some 500 million years ago, Earth's land plants (99 percent of the biomass of Earth) have acted in an intimate dance with the animals and bacterial decomposers to keep Earth's atmospheric gas ratios and temperature remarkably constant. It is the plants who keep temperature constant and who, as they expanded throughout Earth's landmasses, increased the atmospheric content of oxygen from

1 percent to its current level of 21 percent. Without plants, life as we know it would not exist.

Plants breathe in carbon dioxide (CO_2) molecules and during photosynthesis they break them apart; they keep the carbon atoms and breathe out the oxygen. The percentage of oxygen in the atmosphere rose through the plants keeping carbon atoms in their bodies (sequestering). It is kept high through the burial of carbonized plants (charcoal from fires) and the burial of plants as peat, coal, and oil. Only about 0.1 percent of Earth's plants are permanently contained in this carbon "sink," but it is enough to keep oxygen levels high. The rest of the carbon atoms in plants recycle: Any kind of combustion, whether it is logs burning in a fire, bacteria decomposing decaying plants, our bodies slowly burning the foods we eat, or using oil-derived gasoline to power our cars, releases carbon back to the atmosphere as carbon dioxide once again.

High atmospheric oxygen allows rapid chemical processes to occur; all growth accelerates. Animal muscle, which needs a minimum of 10 percent atmospheric oxygen to function, develops. Decaying matter is processed quickly and rocks experience increased bacterial and environmental weathering, which breaks nutrients free for living organisms. For optimum functioning, the ecosystem has to keep oxygen above 15 percent (the point at which fires will burn and large land animals can easily function) and below 25 percent (to prevent uncontrolled raging fires).

Over time, plants have developed specific leaf shapes and leaf positioning to make this process as efficient as it can be. Leaves alternate around a stem or spread out in a flat fan on branches so that those growing above cast a minimum of shading on the leaves underneath. Maximum photosynthesis-capable surface area is exposed. The upper surface of each leaf processes energy from the sun while the underside engages in gas exchange through tiny openings called stomata.

Stomata are essentially tiny lungs. Surrounding them are muscle tissue (much like the diaphragms in our bodies) that constrict and relax, causing the individual stoma to open and close. Basically, plants breathe just like we do. When an individual stoma is full, the opening

All living creatures and plants derive their life from the sun. If it were not for the sun, there would be darkness and nothing could grow—the earth would be without life. Yet the sun must have the help of the earth. If the sun alone were to act upon animals and plants, the heat would be so great that they would die, but there are clouds that bring rain, and the action of the sun and earth together supply the moisture that is needed for life. The roots of the plants go down and the deeper they go the more moisture they find. This is according to the laws of nature and is one of the evidences of the wisdom of Wakan'tanka. Plants are sent by Wakan'tanka and come from the ground at his command, the part to be affected by the sun and rain appearing above the ground and the roots pressing downward to find the moisture which is supplied for them. Animals and plants are taught by Wakan'tanka what they are to do.

—Oku'te (in Densmore, TETON SIOUX MUSIC)

---

closes and carbon dioxide and any other usable molecules are separated from the air. What remains, along with the oxygen generated from the breakdown of carbon dioxide, is released back to the atmosphere as the stoma opens again and breathes out. This cycle is powered by sunlight; at night the plants rest, photosynthesis stops, the stomata are closed. But animal and bacterial respiration and fires do not stop when sunlight is not available; they continue to produce carbon dioxide. Earth's levels of carbon dioxide rise at night and lower during the day—Earth's breathing on a 24-hour cycle. Over a hundred billion tons each of oxygen and carbon dioxide are cycled this way each year.

To keep their airways moist, plants transpire: they take up, or hydraulically lift, water from deep in the ground and breathe it out when they exhale. On a hot summer day, a mature cottonwood tree can breathe out 100 gallons of water an hour. It is so much cooler under a tree or in a forest not so much from the shade cast by the trees' leaves, but from the incredible amounts of moisture that the trees are exhaling. Forests breathe out so much water vapor that from space it is actually possible to see the rain forest creating the clouds that precipitate later as rain. Forests help cool Earth by keeping the air moist, by making clouds, by making rain.

Hydraulic lifting goes on 24 hours a day. At night, when their stomata are closed, the trees, and all deep-rooted plants, deposit the water they are bringing up just under surface of the soil. Some they will use for transpiration the next day but about two-thirds is used by neighboring plants as their primary water supply. Trees literally water their community. Whenever forests are removed—sometimes only half a forest has to be cut—the air and soil begin to dry up, rain becomes scarce, fires are more common, and the land starts to become desert.[1]

The Dance of Plant Chemistry

The carbon atoms that become available from the breakdown of CO_2 during photosynthesis form the backbone of all plant chemistry. Plants use this carbon (along with hydrogen, oxygen, and nitrogen) to make their physical structure (whether a huge redwood or a tiny violet growing along a mountain path); primary compounds such as sugar, starch, and chlorophyll; and hundreds of thousands, perhaps millions, of other, complex, secondary compounds: "acids, aldehydes, cyanogenic glycosides, thiocyanates, lactones, coumarins, quinones, flavonoids, tannins, alkaloids, terpenoids, steroids" and more.[2] Adding to the complexity, all these compounds can be made using different metabolic pathways—different construction techniques, as it were—and each family of secondary metabolites can contain incredible numbers of substances. Simply altering the relationship between four sugar molecules, for instance, can create more than 35,000 different compounds. More than 10,000 alkaloids, 20,000 terpenes, and 8,000 polyphenols are known. About one new alkaloid is identified each day.[3]

Even though many of these compounds are present only in parts per million or even parts per billion or trillion, they exert significant bioactivity. Their bioactivity can increase substantially, sometimes by several orders of magnitude, when they are combined.[4] Through complex feedback loops, plants constantly sense what is happening in the world around them and, in response, vary the numbers, combinations, and amounts of the phytochemicals they make.

The Plant as Seed

Plants put into their seeds the unique chemistries necessary for them to grow when they are released into the world. They also put a large number of compounds in the seeds themselves or on their seedcoat (certain tannins, alkaloids, lactones, phenolic compounds, and flavonoids) to protect them from soil microorganisms. As a result, some seeds can remain viable in soil for years or decades until they sense the right conditions for germination.[5] These chemical mechanisms can be quite sophisticated. *Datura stramonium* (also known as jimson weed),

---

The Chemistry Found in a Single Yarrow Plant

(*Achillea millefolium*), as of 1992

8-Acetocyartabsin
Acetylbalchanolide
Achiceine
Achillin
Aconitic-acid
Alumenum 34
8-Anelooxyartabsin
Apidenin
Apigenin-glucoside
Arabinose
Artemitin
Ascoric acid
Ash
Asparagine
Austricin
Azulene
Balchanolide
Benzaldehyde-cyanhydringlycoside
Betaine
Betonicine
Borneol
Bornyl-acetate
Butyric-acid
Delta-cadinene
Caffeic-acid
Calcium
Camphene
Camphor
Carbohydrates
Beta-carotene
Caryophyllene
Casticin
Cerotinic-acid
Chamazulene
Chamazulene-carboxylic-acid
Choline
Chromium
Cineole
1,8-Cineole
Cobalt
Copaene
Coumarins
Cuminaldehyde
P-cymene
Deacetlylmatricarin
Cis-dehydromatricariaester
Trans-dehydromatricariaester
2,3-Dihydroacetoxymatricin
Dulcitol
Eo
Eugenol
Farnesene
Fat
Fiber
Folic-acid
Formic-acid
Furfural
Furfuryl-alcohol
Galactose
Guaiazulene
Heptadecane
Homostachydrine
Humulene
Hydroxyachillin

for example, coats its seeds with hyoscyamine and scopolamine. These two alkaloids protect the seed from microorganisms and also prevent its germination. When rainfall sufficient for germination leaches the compounds from the seed coat, the seed begins to sprout. The alkaloids disperse through the soil, where they are still strong enough to inhibit other plants and microorganisms in the immediate vicinity.[6] Parent plants often vary the amounts and types of chemicals they deposit in their seeds so that each seed will possess slightly different compounds in slightly differing quantities, helping ensure its survival.[7]

Though dormant, seeds are constantly analyzing the makeup of

5-Hydroxy-3,6,7,4'-tetramethoxyflavone
Inositol
Inulin
Iron
Isoartemisia-ketone
Isobutyl-acetate
Isohamnetin
Isovaleric-acid
Kilocalories
Leucodin
Limonene
Luteolin
Luteolin-7-glucoside
Magnesium
Manganese
Mannitol
Menthol
Millefin
Millefolide
Moschatine
Myrcene
Myristic-acid
Niacin
Allo-ocimene
Oleic-acid
Palmitic-acid
Pentacosane
Phosphorus
Alpha-pinene
Beta-pinene
Ponticaepoxide
Potassium
Proazulene
Protein (1st)
Protein (2nd)
Quercetin
Quercetin-glycoside
Quercitrin
Resin
Riboflavin
Rutin
Sabinene
Salicylic-acid
Selenium
Silicon
Beta-sitosterol
Beta-sitosterol-acetate
Sodium
Stachydrine
Stigmasterol
Succinic-acid
Tannin
Alpha-terpinene
Gamma-terpinene
Terpinen-4-0l
Terpinol
Thiamin
Alpha-thujone
Tin
Tricyclene
Trigonelline
Viburnitol
Water
Zinc

—James Duke,
HANDBOOK OF . . . GRAS HERBS

their external environment through the same complex feedback loops that all plants use. At the moment they sense that just the right conditions exist for them to germinate, seeds begin to release unique combinations of plant compounds such as abscisic and gibberellic acid, cytokinins, and ethylene, which regulate germination and are bioactive at extremely low concentrations—less than 10 ppt.[8] Each seed also contains enough sugar (in the form of starch) to fuel its growth until it can begin photosynthesis on its own. At germination the seed releases an enzyme that begins converting the stored starch to sugar. Each plant uses information from tightly coupled feedback loops between itself and the environment to determine the composition and release of all these chemistries.

A newly developing plant embryo, unlike a human embryo, has no sterile womb in which to grow and so, in a sense, makes its own. As soon as germination begins, the new plant starts releasing compounds through its tiny root system to essentially make a sterile zone around the emerging rootlet. This action protects the seed from harmful organisms and makes space in the soil for its growth. *Solidago altissima* (a species of goldenrod) and *Erigeron annus* (white-top fleabane), for example, release combinations of six to ten different matricaria and lachnophyllum esters (ME and LE) to reduce the growth of plants nearby and so make room for themselves in the soil. Like many phytochemicals these particular esters are highly active at very low concentrations. Dehydromatricaria ester (a type of ME) and LE cause a 50 percent growth inhibition in nearby plants at only 10 to 20 parts per million (ppm).[9] These compounds have a half-life of one to two days before they

---

[Man] sees the morning as the beginning of a new day; he takes germination as the start in the life of a plant, and withering as its end. But this is nothing more than biased judgment on his part. Nature is one. There is no starting point or destination, only an unending flux, a continuous metamorphosis of all things.

—Masanobu Fukuoka, The Natural Way of Farming

biodegrade and must be regenerated through photosynthesis. Still, they are present long enough and in enough concentration that young seedlings can generate a "zone of protection" around themselves until they sprout. All new seedlings have their own unique types of compounds to help them grow. Their environmental feedback loops tell them what chemicals to release, in what combination, and in what quantity.[10]

Because of the diversity of life a new seedling can encounter, its seed chemistry must be able to affect a wide range of fungal, microbial, and plant metabolisms and structures: stomatal function, cell walls, membranes, mitochondria, chloroplasts, chlorophyll production, nuclei, nucleoli, cell division, photosynthesis, respiration, protein synthesis, lipid (or fat) synthesis, enzyme formation, mycorrhizal binding, nodulating bacteria, mineral uptake, and water uptake, among others.[11] The chemicals that are generated can be quite elegant and specific in their actions. For instance, certain seed-released phenolic acids and flavonoids depolarize the electrical potential difference across cell membranes while at the same time altering the permeability of the membranes, thus affecting mineral uptake by the cells of the roots and, as a result, the growth of the plant.[12] Juglone, created by walnut trees and their seedlings, can completely inhibit respiration and photosynthesis in nearby plants and microorganisms.[13] And normally non-water-diffusible lipids can be combined with other compounds to diffuse them through soil, affecting plant germination and growth.[14] *Zapoteca formosa,* a plant that grows in northern Brazil, for example, emits a complex mixture of compounds at germination: nonprotein amino acids, djenkolic acid, taurine, 1,2,4,6-tetrathiepane, dimethyldisulfide, 2,4-dithiopentane, benzothiazole, and a large number of unidentified sulfides and nonsulfur compounds. The root exudate is so strong that a single germinating seed in a petri dish will completely inhibit the germination of 50 lettuce seeds, 25 tomato seeds, or 25 *Acacia farnesiana* seeds. *None* of the compounds show any inhibitory activity alone, but when combined in exactly these ratios they produce one of the strongest germination inhibitors known.[15]

Whatever the mode of action used by the plant the result is the same: space in the soil for the new seedling to sprout.

It is a rare exception when a single substance is responsible for allelopathy [the positive or negative impacts on neighboring plants through the release of secondary plant chemistries]. Isolation of several compounds and different classes of compounds from a particular allelopathic situation has been the consistent pattern. Experiments indicate that either additive or synergistic inhibition may occur from combinations of terpenoids, benzoic acids, organic acids, derivatives of cinnamic acid, p-hydroxybenzaldehyde with coumarin, and the three-way concert action of a flavonoid, coumarin, and phenolic acid.

—Frank Einhellig, "Mechanisms and Modes of Action of Allelochemicals"

The Rhizosphere

Once the seedling has space, breaks through the soil, and begins photosynthesis, it begins to create and release an entirely different set of chemicals into the zone around the roots—the rhizosphere. Organic and inorganic compounds: sugars such as fructose, glucose, ribose, and so on; as many as twenty-seven different amino acids and their compounds; nucleotides; flavones; enzymes; phenolic acids; and more. These may be gaseous volatiles, water-soluble compounds, or more intractable (usually) nondiffusable compounds such as lipids (fats), mucilages, gums, and resins. (This combination of compounds from the roots is very similar to the nectar produced in the plant's flowers.)[16] For every water-soluble compound released by a root, three to five non-water-soluble compounds and eight to ten volatiles are released. Hundreds of compounds, in thousands of combinations, can be involved through thousands of different release sites on the roots.

Plant roots, though confined in space, have an extremely large surface area. A single rye plant, for example, has more than thirteen million rootlets with a combined length of 680 miles. Each rootlet is covered with root hairs, some fourteen billion of them, with a combined length of 6,600 miles. This entire root surface releases differing amounts of chemicals at different locations, strongly regulating the local biocommunity throughout the life of the plant.[17] These compounds promote the growth of bacteria and fungi, stimulate soil microflora respiration, stimulate the growth of nitrogen-fixing bacteria, and increase the numbers and mass of nitrogen nodules, their hemoglobin content, and more.[18] But their actions are not random, nor is it random bacteria, fungi, and microflora that are affected.

Specific bacterial species, like those in the human GI tract and on human skin, have formed symbiotic associations with plant species that have lasted for millions of years. The newly germinating plant releases compounds that literally call the proper bacteria to the area where it is growing. Bacteria are so attuned to these chemicals that they respond to them even if they are present only in parts per billion.[19] As with the human skin and GI tract, once these bacteria cover the surface of the root and fill the rhizosphere, pathogenic bacteria have little room to grow. And these plant bacteria are species specific. Taking two plants with differing symbiotic bacterial populations at random and replanting them in each other's location causes the bacterial colonies present in the soil to completely change. In a short period of time, they will match the composition before transplanting.[20]

> *Working with sorghum, the scientists have also learned that the amount of chemical signal produced by various strains of a single host species can differ as much as a billionfold [and still produce a response].*
>
> —William Agosta, Bombardier Beetles and Fever Trees

The compounds that plants release into the soil are in such combinations and ratios that the health of their bacterial community is maximized. The bacteria respond in kind. Some of the microorganisms, for example, provide metals from the soil, such as zinc, that the plant needs to grow. *Azotobacter* bacteria (and others) have been found to produce plant growth regulators such as cytokinins, which cause increased growth in the plants with which they are associated.[21] *Rhizobium* bacteria form symbioses with legume plants, forming nitrogen nodules on their roots which the plants need for growth.[22] And as with human and animal bacterial symbionts, plant bacteria release compounds that are specifically antibacterial to pathogenic bacteria that threaten the plant. These plant/bacterial populations have been interacting in this way for anywhere between 140 and 700 million years. The better the bacterial population establishes itself and the more healthy and active it is, the healthier and better plants grow.

The newly photosynthesizing plant also releases compounds that initiate the growth of coevolutionary fungi (mycelia) around its roots. The compounds chemically cue specific mycelial spores to germinate, potentiate their growth, and exert powerfully attractive forces on all (already growing) symbiotic mycelia in the vicinity, calling them to the newly emerging plant. As with bacterial cues, these compounds can be strongly active, even at dilutions of parts per billion.[23] The mycelia, which are small hairlike filaments growing throughout the upper layers of the soil, attach themselves to the surface of the seedling's root (as do some bacteria), sometimes penetrating the root body, to form a complex symbiotic relationship called mycorrhiza. This will last throughout the life of the plant. The plant, through photosynthesis, creates sugars and secondary compounds that the mycelia need for growth. The mycelia in turn provide substances the plant needs (nitrogen and phosphate uptake, for instance, can increase 7,000 percent), produce complex polysaccharides that stimulate the plant's immune functioning, and facilitate chemical communication (through its mycelial network) between all the plants in its area. Like symbiotic bacteria, the mycelia generate compounds that protect the plant from pathogenic fungi that attempt to move into the area and harm it.[24] Plants such as Douglas fir may be in symbiosis with mycelia from as many as forty different species of fungi. Mycelial mats, connecting all the plants in a local ecosystem, will sometimes cover hundreds of acres just below the surface of the soil. From time to time, the mycelia will send up their fruiting bodies that we call mushrooms and spread billions of spores.[25]

Everything coming into a plant through its roots and everything going out of the roots has to move through the rhizosphere. An extremely complex relationship exists between the plant and the microflora and microfauna in the rhizosphere region. Sophisticated biofeedback loops, in both directions, carry information that shapes plants'

As the mycelia grow, they constantly encounter tree roots. If the species combination is the right one, chemical signals spark and a remarkable biological event unfolds. Fungus and tree come together to form mycorrhizae, a symbiotic partnership that allows both to benefit. The tree provides the fungus with sugars created from sunlight. The mycelia in turn enhance the tree's ability to absorb nutrients and water from the soil. They also produce growth-regulating chemicals that promote the production of new roots and enhance the immune system.

—Wade Davis, RAINFOREST

---

chemical production. Many of the low-molecular-weight compounds created by plants are exuded into the rhizosphere where they are transformed by rhizosphere organisms into more complex, high-molecular-weight compounds such as polymers. Only then do they become active.[26] Many of these modified compounds combine in the soil with other plant and soil microorganism compounds to form humic acid, one of the most important elements of ecosystem regulation and soil fertility.[27] Thus, soil health is directly dependent on the rhizosphere community and the secondary plant compounds created during photosynthesis. This rhizosphere community is also paramount in maintaining the health of the bacterial underpinnings of the life web. Plant compounds released in sophisticated complexities maintain this health, linking the rhizosphere with the sun through plant actions. The plant, as mediator, increases and decreases the kinds and amounts of phytochemicals produced in order to maintain the optimum levels of health of the soil biota. This dynamic in turn ensures its own health and maximum growth.[28]

Plant Compounds as Medicines for the Plant Itself

As plants grow, they produce a complex assortment of compounds to maintain and restore health. These include: tannins; antibiotic, antimicrobial, and antifungal compounds; mucilages, gums, and resins;

People call the soil mineral matter, but some one hundred million bacteria, yeasts, molds, diatoms, and other microbes live in just one gram of ordinary topsoil. Far from being dead and inanimate, the soil is teeming with life. These microorganisms do not exist without reason. Each lives for a purpose, struggling, cooperating, and carrying on the cycles of nature.

—Masanobu Fukuoka, THE NATURAL WAY OF FARMING

---

anti-inflammatory compounds; analgesics; and so on. They are stored in different parts of the plant, being released in varying combinations and strengths as needed. Often these compounds are highly reactive when combined or exposed to air and so are kept isolated in holding cells located throughout the plant. The plant can increase the quantity of any of these compounds at the point of need or translocate them extremely rapidly through its tissues.[29]

Antifungal, antibiotic, or antimicrobial (preinfectious) compounds protect the plant from invading pathogenic organisms. For example: The tulip tree *(Liriodendron tulipfera)* produces a number of strongly antimicrobial alkaloids (dehydroglaucine and liriodenine) that it stores in its heartwood to protect it from invasion by microorganisms. Chicory *(Cichorium intybus)* produces a number of strongly antifungal compounds to protect its leaves and roots from pathogenic fungi. The compounds are so potent that even when chicory roots are kept moist on a plate for lengthy periods they will not mold. Other chicory compounds strongly protect against damage or infection from nematodes and other small organisms.[30] Plant antimicrobial compounds such as those in chicory are active against microorganisms in exceptionally minute concentrations, ranging from one part per thousand to one part per million.[31] During infection other kinds of compounds can be brought into play. Aromatic coumarins in such plants as potatoes increase rapidly at the site in response to any pathogenic organism.[32] Cyanogenic compounds are also commonly present in at least a thousand plants where they are released as hydrogen cyanide gas to kill invading organisms.

In many instances invading pathogens release their own compounds that are toxic to the plant. Plants immediately begin to identify these compounds and create chemistries designed to counter them. At the same time, the plant will begin to generate unique compounds—phytoalexins—at the site of infection that are never present in the plant until an infection occurs. When fungal spores take hold on a leaf surface, for instance, and begin inserting growth tubes into the leaf, a plant may begin to synthesize a phytoalexin specific for that fungus. The synthesis begins immediately, can be detected after an hour or two, and reaches its highest concentration in 48 to 72 hours. The phytoalexin is concentrated in leaf cells and pushed out onto the surface of the leaf where the fungus has taken hold.[33]

Many plants also possess tannins; the amount and type of tannins in different kinds of plants covers a broad range. Tannins are held throughout the plant—in buds, leaves, roots, seeds, and stems—and only become active during cell breakdown. Tannins perform a wide range of actions. They have mild bactericidal properties and act as a barrier against penetration and colonization from plant pathogens, and they are astringent. In other words, tannins dry out or diminish the leakage of fluids from any break in plant cells and cause contraction of the tissues. They also bind with plant proteins and polysaccharides on the surface causing a hardening (or "tanning") of the cells. This creates a protective surface layer that reduces the ability of toxins or pathogens to gain entry to the damaged area. The strengthened cellular surface also protects the tissues against external irritations and is highly inhibitory to a wide variety of viruses.

When cell breaks occur some plants will also release complex polysaccharides—gums or concentrations of mucilaginous compounds—into the area. Gums immediately lower the sugar concentration in the damaged area, decreasing yeast and bacterial feeding; directly interfere with yeast feeding and bacterial growth because of their indigestibility; absorb toxins and wastes to move them away from the area; and soothe and protect the irritated tissues of the plant just under the tannin surface hardening.

Resins, or complex combinations of terpenes, are released by some plants, such as evergreen trees, during cellular breakage. They are antimicrobial, antifungal, and antibacterial, acting against a wide range of organisms. Their flow into the wound in liquid form seals the cavity and then hardens, protecting it from further damage.

Significant numbers of anti-inflammatory compounds, such as quercetin, are also made by plants, and are released into damaged areas to control cellular inflammation. Salicylic acid, a precursor in the production of aspirin, is extremely common in plants (from whom it was originally isolated). Like most plant chemistries, salicylic acid possesses scores of uses depending on dosage, location, and combination. In low doses, released into the ecosystem, it stimulates plant growth, but in larger doses it inhibits growth. It is also an anti-inflammatory compound that plants can translocate rapidly to wherever it is needed.[34] It is often used in series with jasmonic acid (JA).

JA, which is produced by many plants when cells are damaged, triggers inflammation and plays a significant role in stimulating phytoalexin production. Increased dosages interfere with insect digestion, and in large quantities JA hastens plant aging, stimulating the plant to produce seeds before it dies. To regulate the degree of inflammation plants increase production of salicylic acid in the area of the wound or move it to the area through vascular transport. Salicylic acid blocks the production of inflammatory compounds such as JA and so decreases inflammation.[35]

Because plants work so intimately with oxygen as they break apart CO_2 and H_2O molecules, special problems arise. Oxygen is highly mutagenic (causing mutations in genetic structure), carcinogenic (causing cancers), and oxidative (causing cell damage from oxidation). In consequence, Earth's plants produce thousands of compounds that possess antimutagenic, anticarcinogenic, and antioxidative actions. Such compounds are rapidly translocated throughout plants, constantly regulating the effects of oxygen on living tissues.[36] Like most plant compounds, they also have multiple uses: Plant compounds that inhibit cell division may be used as anticancer compounds or to deter cell division in seedlings attempting to germinate within another plant's zone of protection.[37]

Protection from Animal Predation

Plants also generate hundreds of compounds that they use to protect themselves from being overconsumed by insects and animals.[38] While some plants use these compounds to reduce *all* foraging, most plants tolerate, or even enjoy, about an 18 percent foraging rate (10 to 25 percent depending on the plant) before they begin to initiate high levels of protective compounds.[39] Plants and animals have coevolved over a long period of time and their relationships reflect mutualistic interdependencies that are often millions of years old. Many of the actions of animals when they eat plants (termed "herbivory") are necessary for both plant and ecosystem health. Herbivory alters the density, composition, and health of plant communities through eating plants, dispersing seeds, and defecation. Some plants produce an initial series of leaves designed to be eaten, and more luxurious growth only occurs once that has happened.[40] For many plants, metabolism, respiration, and metabolite transport are all stimulated by animal and insect feeding.[41] It is only after foraging rises above a certain level that many plant defensive compounds are produced in quantity or come into play.

The compounds that plants use to deter high feeding levels may occur in a number of forms. They may be bioactive glycosides, which are toxic compounds connected to sugar molecules. These only become active when the sugar molecule is consumed or detached. They may be combinations of separate compounds kept isolated in leaf structures that only come into contact with each other during herbivory. They may only react when exposed to oxygen if the leaf is crushed or chewed. Or they may be compounds that have specific and powerful effects on the internal body chemistry and organ function of insects or animals if they are taken in quantity.

Glucosinolates, which occur in plants such as mustards and horseradish, only become irritants to tissues when plant tissue is crushed or chewed. An enzyme is sequestered in the plant that, during chewing, combines with the glucosinolate to produce volatile, burning, mustard oils. Cyanogenic compounds that apples, cherries, and at least a thousand other species of plants make, are sequestered in a like fashion in plant cells, often in their seeds. The hydrogen cyanide gas that forms during cell crushing, among other things, inhibits the respiration of any foraging organism (which is why it works against bacteria).[42]

Some of the plant compounds, such as tannins, produce negative effects simply from the amount ingested. Though some tannins are tolerated in the diet, tannins in any quantity are commonly avoided by most foraging animals. (Those animals that do commonly eat quantities of tannins often show exquisite sensitivity to the exact chemical makeup of the tannins present in any species of plant.)[43] All tannins, though some more strongly than others, interfere with the ability of animal stomachs to process the plants as food. Tannins combine with proteins ("tanning") to form chemical bonds impervious to breakdown by animal and insect digestive enzymes.[44] The proteins and tannins in plant leaves are compartmentalized. Chewing by insects or animals breaks the compartments open and combines the two compounds, rendering the protein unusable. Most animals or insects eating a tannin-heavy plant diet will starve.

Other compounds, like some sesquiterpene lactones found in plants such as feverfew, yarrow, and blessed thistle, are strongly antimicrobial and interfere with the microbial composition of large animals' rumens (or forestomachs). This initial stomach in herbivores contains large microbial populations that digest plant materials before they are sent to the secondary ("regular") stomach for processing. The rumen is in essence a large liquid fermenter filled with a massive microbial population. The plants the animal eats feed the microbes whose bodies supply much of the protein the animal needs to live. Without this microbial population herbivores would die. Strongly antimicrobial compounds, eaten in large quantities, can kill off segments of this population, thus affecting the animals' metabolic functions. (Tannins may also bind with bacterial proteins in the rumen, again interfering with protein uptake and affecting nutrient levels.)

More rarely, plant compounds can be potent neurotoxins if taken in quantity and can cause a syndrome similar to Parkinson's disease in animals such as horses. Solanum-type steroidal alkaloids found in such plants as potatoes or the silver nightshade *(Solanum eleagnifolium)* that grows in the Sonoran Desert among ironwood trees are cholinesterase inhibitors affecting the nervous system. Animals ingesting large quantities (or sometimes smaller amounts from plants that produce them in more potent forms) can experience apathy, drowsiness, salivation, labored breathing, trembling, ataxia, and muscle weakness. At higher doses such compounds can cause convulsions, paralysis, unconsciousness, coma, and death. (It is important to note here that no animals in the wild have been found to *ever* eat quantities sufficient to cause death. The initial symptoms of apathy, ataxia, and muscle weakness, from which they readily recover, cause any foraging animals to stop eating long before death occurs. It is only in laboratory settings or certain overcrowded farming situations, where the animals are forced to eat unlimited quantities, that they will do so until they die.)

The secondary compounds in plants vary by species, season, time of day, individual plant, and environmental stressors. Plants continually receive and process information from their environment and they use it in determining the amounts and types of secondary compounds they need to produce and in what combination. When being overeaten by aphids or caterpillars some plants quickly combine and release a complex mixture of imino acids and sulfur amino acids. Relatively inactive in pure form, the combined chemicals strongly deter feeding at concentrations ranging from one part per thousand to one part per million.[45]

Plants can also increase their production of many compounds at need. Plants growing in areas where there is little animal feeding produce very few antifeedants. However, if animals move into their area and begin heavy foraging, or if the plants are transplanted to an area where such foraging occurs, they will begin producing these kinds of compounds very quickly.[46] Clover (*Trifolium repens*), as an example, when growing in Russia, doesn't produce cyanogenic compounds; in Britain it does. In Russia the temperature drops enough each winter to significantly reduce the population of the snails and slugs that feed on the plant, thus keeping foraging levels low. In Britain the temperature never gets cold enough to reduce the snail populations and the plants begin production of cyanogenic compounds to reduce foraging levels.[47] Tannin-producing plants that are experiencing predation above tolerable levels can increase tannin content fairly quickly, rendering plant proteins unusable as food. Such chemical increases can be dramatic: Scopoletin content in individual oat seedlings can increase from 3.8 micrograms to 121.9 micrograms in response to environmental stressors.[48] And, in response to Colorado beetles, tomato and potato plants release a proteinase inhibitor inducing factor (PIIF) into their sap, which quickly circulates throughout the plant. The proteinase inhibitors produced in response to the release of PIIF can comprise 2 percent of plant protein within 48 hours. Proteinase inhibitors, like tannins, inhibit the enzymic digestion of plant proteins.[49]

Douglas fir trees show a sophisticated chemical response to infestations of spruce budworm. The trees release a complex mixture of volatile oils, or terpenes, from their needles. This includes *a*-pinene, camphene, *B*-pinene, myrcene, limonene, carene/*a*-terpinene, terpinolene, bornyl acetate, ceironellyl acetate, geranyl acetate, thujene, linalool, cadinene, sesquiterpenes 3 and 4, sabinene, ocimene, *y*-terpinene, terpinen-4-ol, *a*-humulene, and a number of other, unidentified terpenes.[50] Douglas fir trees regulate the amount of individual and total terpene production depending on time of day, year, location, and predation. The trees increase terpene production 500 to 600 percent between June 13 and July 29 in the western United States and continue increasing production more slowly until late September. This exactly parallels budworm activity and reproduction. One component, boryl acetate, which is strongly toxic to budworms, increases in quantity from about 0.25 mg/g to 1 mg/g (fresh weight) between June 18 and July 8. The female and male spruce budworm respond differently to different terpenes and the trees vary them accordingly; evenness of terpene production in a tree leads to increases in budworm populations. Trees that vary the production of differing terpenes (such as terpinolene, which can attract female budworms if present in too high a quantity) affect the ability of the budworm to find trees, to feed and reproduce. The trees *vary the composition and production of terpenes each year* thus decreasing the ability of the budworm to develop widespread immunity to specific compounds. The terpene composition that determines strong resistance to predation varies each year. However, *these kinds of variations and increases are only seen when a budworm population is beginning to increase in numbers.*[51]

Plants can also affect the reproduction of foragers that feed too heavily on them. Chemicals produced by *Calliandra* species lower sexual activity and offspring survival in many insects, including aphids, that overfeed. Many of the compounds are not toxic until up to five days after feeding and are not intended to deter individual feeding but to reduce species abundance and collective impact over time.[52] Other plants produce compounds that are identical to insect hormones which, in tiny quantities, can interfere with the developmental stages of insects who eat it. They produce abnormalities in growth, sterility, and early death.[53] Most plants, such as the coastal fir tree *(Abies grandis)*, produce these compounds only in response to infestation.

This kind of activity is fairly common, especially by plants that produce phytoestrogens or plant compounds that mimic the action of animal estrogens. Female rats that consume too much estrogen-containing plant matter experience ovulation disturbances[54] and certain isoflavonoids found in alfalfa and ladino clover (and other plants) mimic progesterone and can cause infertility, reduced lactation, and difficult labor.[55] Significant sheep infertility problems often occur for Australian ranchers with large herds. A main foraging plant, subterranean clover *(Trifolium subterraneum)*, produces estrogenic compounds that affect sheep reproductive cycles and cause up to a 70 percent reduction in ewe fertility. The compounds are a mixture of two isoflavones, genistein and formononetin, which are active analogues of female estrogen. They make up about 1 percent of the plant. Coumestrol, a similar substance found in alfalfa and ladino clover, is 30 times more active than the compounds in subterranean clover. Like many plant estrogens an animal has to eat it in quantity to severely

Because plants are firmly rooted in the soil and cannot run away from their enemies, they have long been considered passive in interactions with other organisms. This view has been falsified by several decades of research on plant-pathogen and plant-herbivore interactions.

—Marcel Dicke, "Plants in Action"

Chemical communication between herbivores and plants is the primary driving force behind what we recognize as coevolution, and that process is better represented as an accelerated process of genetic change facilitated by the efficiency of chemical information exchange.

—Kevin Spencer, "The Chemistry of Coevolution"

---

reduce fertility. It is converted by animal metabolism into a compound identical to that isolated from pregnant mare's urine in the making of the estrogenic pharmaceuticals used by women. One million Australian ewes fail to produce lambs from eating such plants each year.[56] The plants vary the amount of these compounds by year depending on a number of environmental factors including density of foragers, essentially regulating the population levels of herbivores.

Chemical Communication from Plants

In response to overfeeding by aphids some plants will release a volatile aromatic, *E*-beta-farnesene, from their leaves. This mimics an aphid alarm pheromone warning of approaching predators, telling them to flee the plant.[57] Spider mite–infested lima beans will release a blend of volatile oils (terpenoids) that attracts a predatory mite that feeds on spider mites. The plants can tell exactly what kind of mite is feeding on them through analyzing the chemistries of their saliva. Each plant species then produces a different blend of volatiles depending on what kind of spider mite is feeding on it. That mix will *only* call the predator that feeds on that kind of mite.[58] The plants also tell other, uninfested, lima beans what is happening. Those receiving the communication also begin to release the chemical that calls predatory mites, thus reducing the spread of the feeding mites.[59]

Many plants who detect a component in caterpillar saliva [*N*-(17-hydroxylinolenoyl)-L-glutamine] called voicitin, begin emitting a blend of volatile aromatics (a combination of 6-carbon aldehydes, alcohols, esters, indole, several terpenes and sesquiterpenes). This calls wasps who are parasitic on the caterpillars. The components are emitted only if feeding goes on longer than two hours, and, like spider-mite-infested plants, the whole plant emits the compounds, not just the area that is damaged. The amount of volatiles released varies with time of day, being particularly high during the period when the wasps forage.[60] Like the production of plant estrogens this chemical signal to the caterpillar predators is designed to reduce caterpillar populations, thus reducing levels of predation in succeeding generations. (If the caterpillar quits feeding, the plant immediately stops producing the compounds.) When parasitoid wasps detect the chemical signal, they locate the plant and deposit their eggs in the caterpillar. In some instances the plant also begins to generate plant compounds that slow the caterpillar, making it more available to the wasp.[61] Other plant species have developed close mutualistic relationships with ants to protect themselves.

In Central America and Africa certain species of *Acacia,* a large shrub or small tree, is covered with thorns, some of which are hollow and house ants. Much like coevolutionary bacteria, *Pseudomyrmex* ants recognize new shrubs as coevolutionary partners and colonize them. The trees produce special nectar along the stems for the ants to eat. Like the compounds released from plant roots, this nectar contains a rich mix of fats (lipids), proteins, sugars, and other compounds necessary for the ants to remain healthy. The ants remove vegetation from around the base of the plant, remove leaves of other plants that shade the tree, kill any vines that try to grow up the tree, and attack any herbivore that tries to eat the plant.[62]

South American leafcutter ants collect plants, chop them up, and

---

It is those who believe only in science who call an insect a pest or a predator and cry out that nature is a violent world of relativity and contradiction in which the strong feed on the weak. . . . These are only distinctions invented by man. Nature maintained a great harmony without such notions, and brought forth the grasses and trees without the "helping" hand of man.

—Masanobu Fukuoka, The Natural Way of Farming

feed them to a fungus that they grow for their food. When forming new colonies the ants transfer starters to the new colony—somewhat like a sourdough starter handed down for generations. The fungus the ants grow can sometimes become infected by an *Escovopsis* microfungus. This fungus is kept in check by a *Streptomyces* bacteria that is symbiotic with the ants and grows on their bodies. The *Streptomyces* also produces growth compounds that significantly increase the biomass of the fungus; the ants apply the substances made by the bacteria to the fungal colony to maintain its health. Ants have been living in close mutualistic relationships with acacia, their fungus gardens, and their symbiotic bacteria for at least 50 million years.[63]

Plant Chemistry and the Soil

After a plant disperses its seeds, in the colder latitudes, the seasons begin to change. Trees and other perennials begin pulling chemistries out of their leaves, storing them in their trunks and roots for future use. About half of the leaf chemistry is retrieved; the leaves begin to change color in response.[64] The annuals die altogether, having sent their life onward in their seeds. All these dying leaves and plants come to rest on and in the soil, creating a thick blanket of plant matter. Not all this plant "litter" is dieback, of course; a consistent portion is living leaves, limbs, lichens, and trees that for one reason or another fall to the ground. "Living litter" is much more highly active in its chemistry than dieback. The trees and plants continue to hydraulically pump up water from deep in the Earth and release it through their stomata, forming clouds, and then it rains.

Rain percolates through the thick bed of plant matter that is resting

---

This use of antibiotics by ants to treat their fungal gardens raises an interesting question, why has Escovopsis *not evolved resistance to the antibiotic? . . . Presumably, part of the solution is that the antibiotics involved are not the standard product of a pharmaceutical company.*

—David Wilkinson, "Ants, Agriculture, and Antibiotics"

on top of the soil and, like water through the coffee grounds in a coffee filter, leaches the chemistries from the plant matter into the soil. The lumpy shape of the Earth allows water to accumulate in standing pools filled with leaves and old plants and twigs and here the chemicals concentrate in strength—like a tea that has been left steeping. All of the plant litter releases its chemistries into the soil at varying rates depending on what kinds of plants it is from and what kinds of compounds are in it.[65] The soil then separates out and routes the thousands of carbon chemistries being deposited into it.[66] Many of these substances, such as flavonoids, degraded lignin, terpenes, lignans, and tannins recombine to form humus—what we primarily think of as dirt (though dirt actually includes as well the billions of other organisms that live within it).

Humus is mostly two substances, humic acid and a combination of polysaccharides or sugar molecules. No one knows how humic acid forms, but once formed it acts like a living substance and possesses a number of unique characteristics. It forms crystals, much like snowflakes in a sense, and, like snowflakes, no identical ones have ever been found. Humic acid "uncouples" many plant compounds, separating them into their constituent chemistries, detoxifying them, and keeping the soil fertile.[67] As well, it stores the separated chemistries it has received within itself as stable complexes where it can, when inputs from the ecosystem indicate the necessity, recombine them into needed compounds and rerelease them into the ecosystem. Humic acid acts, in essence, much like a storage battery for the plants' complex chemistries. As long as the plants are promiscuously producing compounds that regularly fall in a resource cascade to the ground, the battery remains full, the soil rich and bountiful.[68] Through tightly coupled feedback processes information on the chemistry reserves stored in humic acid feeds back into the aboveground plant communities, indicating what plants should grow in what combination in what ecosystem and what kinds of chemistries they should produce to keep the soil healthy.[69] (This is why it is not possible to increase soil fertility through human action. Unless interfered with, the soil in natural ecosystems is *always* at maximum fertility. To increase soil fertility at one location means that ecological resources have to be taken from someplace else.

Soil fertility is temporarily increased at one location by decreasing it in another. This is not even a zero-sum game. The removal of ecosystem resources in that one location causes a diminishment of its functional community, from which it cannot recover except over very long time spans. Earth's plant communities have been doing their job for at least 500 million years; only extreme hubris would lead us to believe we could do better through an agricultural science not even one hundred years old. Mimicry of the natural dynamics of plant communities would seem a better and clearly more sustainable approach.)

Plant pollen also contributes significantly to the chemistry and formation of soil. Each year millions of pounds of pollen are released to the Earth's ecosystems. Wind-pollinated plants such as grasses, ragweeds, alder, birch, poplar, aspen, willow, spruce, and conifer release huge quantities of pollen. Even angiosperms contribute to this stream; wind or animals that brush against them can release tiny pollen rains to the ground in their vicinity. A significant number of streambank trees are wind pollinated and their pollen rain falls in huge quantities into the water, where the constituent chemistries quickly leach out. (A great deal of plant litter also falls into streams, where their constituents also leach out, contributing to stream purity and vitality.)[70] Most ecosystem pollen produces the same kinds of effects in ecosystems as it produces in people who take it medicinally: a nutrient and secondary metabolite cascade into the ecosystem that is taken up exceptionally quickly and sparks a growth burst in response. Pollen is often rich in steroidal compounds such as brassinoloide, a potent growth stimulator.[71]

The leachate from plants and pollens that naturally occur in ecosystems synergistically enhances the growth of plants normally living in those ecosystems and inhibits the germination of seedlings not normally growing there, thus maintaining ecosystem dynamics. The growth stimulation effect on system seedlings from community leachate can be as high as 1,000 percent.[72]

The rates at which the chemistries in litter fall move back into the ecosystem vary depending on the ecosystem involved, from weeks to centuries. During that entire time they modify soil chemistry and soil community composition.[73] Anything that changes the surface plants' chemistries changes the composition of the underlying soil, humic acid, soil community, and local streams. Overgrazing by herbivores, for instance, induces increases in the plant chemistries that cascade into the soil ecosystem, changing its nature.[74] This process explains, in part, why commercial logging impacts so negatively on forest soils. It deposits abnormal amounts of green, highly active plant matter onto the ground. The resin-filled roots remain within the soil, tons of leaves and needles are deposited on top. This releases huge quantities of highly bioactive plant chemistries all at once into local habitats. The combined chemistries are so strong that even after eighty years many forest soil systems and plant communities still have not been able to process them.

Water percolating through dead or dying plant litter is not the only way plants leach compounds into the soil. Rain, fog, dew, and melting snow leach prodigious quantities of secondary compounds from living plants. Rain dripping through the canopy or running down the stem of a tree or the even tinier canopy of a yarrow plant carries hundreds of plant compounds down to the soil. Lichens that grow on trees (and throughout the ecosystem), each possessing their own unique chemistries, also leach substantial quantities of secondary metabolites, which each combine in unique ways with those of their host plants.

Lichens are a symbiosis of two organisms, a fungal mycelia and alga, that come together to form one plant. In a sense they are a kind of mycelia that can live outside the soil; the alga protects the internal mycelial core (from such things as drying out) much as the soil does. The exterior alga and the interior mycelial core possess their own unique biochemistries while, in combination, they produce synergistic compounds unique to their combination as a lichen. In the case of *Usnea* lichens the exterior covering is highly antibacterial while the inner mycelial core, composed of complex polysaccharides, is strongly immune potentiating. Lichens release these compounds into their tree hosts to support their hosts' health and also leach them into the system during rainfall. The tree lichens growing in a hectare (2.5 acres) of oak woods can weigh 1,500 pounds and their chemistries significantly alter throughfall water chemistry.[75] When lichens fall to the ground they

We still have very little understanding of the ecological significance of diversity. If systems are highly integrated with intricate webs of species interactions, diversity may feed upon diversity with so many cross-linkages that prediction of global perturbations will be virtually impossible.

—W.J. Bond, "Functional Types for Predicting Changes in Biodiversity"

---

often take the small twigs to which they are attached with them, adding more mass to litter fall.

Tree shape interrupts the flow of fog across landscapes, causing water condensation onto limbs, trunks, and leaves. The fog runs down into the ground, leaching chemistries in the process. Sixty-six percent of the water used by understory plants in areas of heavy fog comes from tree-chemistry-enriched water or leachate. Leaves on the trees in forests where fog frequently occurs have, over time, developed shapes that facilitate water gathering from fog. They essentially have learned how to collect water from clouds. Their understory plants cannot live without it.[76]

Soil community organisms and insect and animal herbivores also help the movement of plant chemistries into ecosystems.

Mobile Rhizospheres

As bacteria created more complex land life-forms they simply enclosed bacterial communities inside them—basically making mobile rhizospheres. Mammal and insect herbivores, birds, worms, and other soil fauna take in plant matter, process it through their internal, mobile soil communities, and deposit it back onto (or within) the ground as "night soil."[77] These mobile soil communities move throughout ecosystems, intricately intertwined with plant communities. They assist the breakdown and distribution of plant chemistries throughout the ecosystem.

Rather than being pests, insects graze throughout ecosystems, much like herds of buffalo or other herbivores. They move through ecoregions, chew up plant matter, ingest it, remove constituents they need, their intestinal community works on what is left, and they excrete the remainder onto the soil. This moves plant chemistries throughout and between ecosystems. Though few people think about it, insect excrement (frass fall) is a significant contributor to the soil. Insects feeding on eucalypts, for example, can produce up to 5,000 pounds of frass per hectare (2.5 acres) per year. The compounds in eucalyptus leaves, in this instance, are metabolized in the insect gut to form compounds highly inhibitory to the mustard plants that often try to move into eucalyptus territory. Many mammals alter ingested plant chemistries in the same fashion, contributing to plant growth patterns.[78]

Worms are like a herd of buffalo under the soil. Most live within the top eighteen inches of soil but some will go as deep as eight or nine feet. An acre of land can contain from 50,000 to a million worms. They literally pull entire leaves into their burrows and eat them; an acre of worms can eat ten tons of plant litter in a year.[79] Much like earthworms, insects and mammal feeders, bacteria, protozoa, nematodes, microarthropods, millipedes, and so on all form essential parts of the community of the soil. They break up plant litter into smaller bits so that more surface area of the plant is exposed to the leaching effects of water and the actions of soil microorganisms.[80] With greater surface area exposed, plant chemistries are more efficiently leached into the system—basically the difference between dripping water through whole coffee beans or finely powdered ones. Bacterial enzymes break apart many of the complex biomolecules that are left into simpler compounds that can be more easily incorporated into humic acid.[81]

Plant Aerosols

Plant chemistries also move into the ecosystem and soil through their release as aerosols from the surface of plant leaves and, sometimes, stems. Chemical volatiles that are released to guide or call pollinators make up some of this flow, but plants also release huge quantities of terpenes and other aromatics as well. The world's evergreen forests release more than 1,000 megatons (two trillion pounds) of terpenes per year.[82] Additional trillions of pounds of terpenes are released by such plants as yarrows, artemisias, and ambrosias, and these all combine

Might there not be souls which bloom in stillness, exhale fragrance and satisfy their thirst with dew and their impulses by their burgeoning? Could not flowers communicate with each other by the very perfumes they exude, becoming aware of each other's presence by a means more delightful than the verbiage and breath of humans?

—Gustav Fechner (in Tompkins and Bird, THE SECRET LIFE OF PLANTS)

---

with trillions of pounds of scores of other aromatic volatiles. They infuse the atmosphere, sometimes traveling great distances, and fall as a constant rain down onto the soil, plants, and microorganisms beneath the plants that generate them. Plant communities live in clouds of unseen aromatic volatiles. They breathe them in; their bodies are coated with them. These volatiles are significantly bioactive in plant systems; plants can be anaesthetized by gases such as chloroform, just like people.[83] Some of the volatile terpenes that are released, cineole and dipentene for instance, are potent inhibitors of oxygen uptake by mitochondria. This inhibits respiration by bacteria and newly germinating seedlings but does not appreciably affect mature plants. Other volatiles, such as the coumarin esculin and a number of sesquiterpenoid lactones (arbusculin-A, achillin, and viscidulin-C), significantly increase the respiration and stomatal activity of mature plants.[84] (In certain ecosystems the terpenes released by such plants as creosote bush, eucalyptus, certain artemisias and salvias, are so inhibitory that few other plants can grow under their overstory.)[85] Just a few of the actions of plant terpenes: They purify the air, modulate plant emergence, enhance the respiration of the plant community, feed into mycelial networks, and play an essential role in the formation of humic acid.

Plants, and their chemistries, do even more, of course. They are intimately interwoven into the lives of all organisms on Earth. And the roles of plants are still more complex. They exist not for themselves alone; they create and maintain the community of life on Earth, they produce the chemistries all life needs to live, and they heal other living organisms that are ill.

Plants as Medicines for All Life on Earth | 8

Plant people have a way of being invisible,
of blending perfectly with the landscape,
or of being visible only to other plant people. . . .
They are doctors,
though the doctoring of some of them is most subtle.

—Dale Pendell, *Pharmako/poeia*

My elders have said to me
that the trees are the teachers of the law.
As I grow less ignorant
I begin to understand what they mean.

—Brooke Medicine Eagle

Imagine a ball of twine the exact size and shape of Earth. Better yet, telephone line. Take the end point of the line and weave it back into the beginning so that there is no beginning and no end. Every place the line crosses itself (you could think of them as synaptic junctions) messages cross over; communication travels quickly throughout the entire line itself as well. Academic disciplines are areas where a segment of line is cut out of the ball and studied. They explore its tensile strength, its molecular structure, its chemical composition, the colors and types of wires that run through it. Any communications that were flowing or might flow through it cannot be studied once it is cut out of the whole—only a tiny part of the picture can be seen. Misunderstandings easily arise, especially if the communications that flow through the line are the most important thing.

Turn the ball of telephone line back into Earth. Each plant, plant neighborhood, plant community, ecosystem, and biome has messages flowing through it constantly—trillions and trillions of messages at the same time. The messages are complex communications between all the different parts of the ecosystem. There is no beginning and no end, no cause and no effect. The three-and-a-half-billion-year-old feedback loops of Earth are so closely intertwined that there is always another cause underneath whatever cause you begin with. Impacts at any one point affect every other point in the system. Life is so closely coupled with the physical and chemical environment of which it is a part that the two cannot legitimately be viewed in isolation from one another. As James Lovelock says: "Together they constitute a single evolutionary process, which is self-regulating."[1]

This recognition of Earth's self-regulating nature led Lovelock to understand Earth as a living being, not a ball of resources inhabited by human beings hurtling through space. The novelist William Golding, a neighbor, suggested a name to him: Gaia, an ancient Greek name for the living, intelligent, and sacred Earth. And Gaia, at four billion years of age, is very old compared to us. Even the plants are ancient compared to our tiny life-spans, having begun to appear some 700 million years ago.

There is a holly in Tasmania that is 43,000 years old, a creosote

There are men charged with the duty of examining the construction of the plants, animals, and soils which are the instruments of the great orchestra. These men are called professors. Each selects one instrument and spends his life taking it apart and describing its strings and sounding boards. This process of dismemberment is called research. The place for dismemberment is called a university.

A professor may pluck the strings of his own instrument, but never that of another, and if he listens for music he must never admit it to his fellows or to his students. For all are restrained by an ironbound taboo which decrees that the construction of instruments is the domain of science, while the detection of harmony is the domain of poets.

Professors serve science and science serves progress. It serves progress so well that many of the more intricate instruments are stepped upon and broken in the rush to spread progress to all backward lands. One by one the parts are thus stricken from the song of songs. If the professor is able to classify each instrument before it is broken, he is well content.

—Aldo Leopold, A Sand County Almanac

---

bush in the American Southwest that is 18,000, a box-huckleberry up north over 13,000, and a grass colony covering half a square mile 1,000 years old. The oldest Bristlecone pine is nearly 5,000 years of age, the oldest redwood was 3,200 years old when it was cut, there is a bald cypress in Oaxaca, Mexico, that is over 2,000, and some lichens grow so slowly that they only add millimeters to their size every century. The Indian banyan tree (a type of fig) can store 25,000 gallons of water within it and live more than a thousand years. Thousands of pillarlike prop roots rise up to support massive limbs; a single tree can cover acres. Banyans can grow so large that Alexander the Great could camp with 7,000 of his men under a single tree. There is a thousand-year-old mycelial network that covers 1,500 acres in Washington State and another one 1,500 years old in Michigan covering 38 acres. There is an aspen grove whose root system has spread over 106 acres, has lived for 10,000 years, and weighs 6,500 tons. The innocuous herb that is plowed under a housing development can be 200 years old, the osha root picked for medicine 300, the shrub rooted out for a parking lot 800. Judging the actions of these plants, their functions in ecosystems, and their chemistries through the timescale of a human life often misses what can only happen in decades, centuries, or millennia.

It is our temporal limitations that prevent most of us from noticing what plants do over such scales of time. For instance, from recognizing that plants and plant communities possess tremendous powers of movement, that their movement shows intention, that they can cross thousands of miles when motivated, and that their movement patterns are not random but are determined by large-scale feedback loops millions of years old. On short, localized scales: Climbing plants that need support will grow toward a trellis, and if the trellis is shifted the plants will change direction. On long scales this can be even more pronounced, though it is harder to see. As ecologist K.D. Bennett observes: "On long timescales, herbs may be much more mobile than they would appear from observation of processes occurring within the timescales of research grants (or careers), and this is because other processes become involved and are, in fact, the processes that determine [their] overall distribution."[2]

Plants circulate throughout ecosystems, between ecosystems, and across and between continents; the longest seed dispersal distance known (without human help) is 15,000 miles. Plants, in fact, move themselves throughout landmasses and across distances that mere seed dispersal dynamics and mathematics cannot explain. The places

---

At the beginning of the twentieth century a gifted Viennese biologist with the Gallic name of Raoul Francé put forward the idea, shocking to contemporary natural philosophers, that plants move their bodies as freely, easily, and gracefully as the most skilled animal or human, and that the only reason we don't appreciate the fact is that plants do so at a much slower pace than humans.

—Peter Tompkins and Christopher Bird, THE SECRET LIFE OF PLANTS

they move to and the ways that they arrange themselves in ecosystems are not accidental and are not random.[3] Plants arrange themselves in ecosystems and throughout continents to fulfill specific functions; their spatial arrangements exist for a reason.

Conventional Western epistemologies limit conception of what plants can do, and short human attention spans interfere with being able to see plant functions that exist over extremely long cycles and large systems. Most ecological field studies contribute to the problem: They are generally less than three years in length and 95 percent of them occur on plots less than 2.5 acres in size—half of them occur in a 9-square-foot area or smaller. Few of the researchers have a personal, long-term relationship with the area they are studying.[4] Such difficulties of scale and time are compounded in a number of ways. One is the language we use to name plants, the Latin binomials by which they are classified.

The system of naming plants created by Linnaeus gives plants a genus and species name. For instance, osha is named *Ligusticum porterii*—the large grouping or genus is *Ligusticum* and the kind of Ligusticum, or species, is *porterii*. *Ligusticum* means "of Liguria," referring to a geographical area of Italy; *porterii* means "of Porter" or "Porter's," referring to a Philadelphia botanist, Thomas Conrad Porter (1822–1901), who roamed over the country naming plants after himself. So this uniquely North American plant interwoven into local ecosystems for hundreds of thousands of years now has as its name "Porter's Liguria, Italy." And except for a number of European plants (such as *Achillea millefolium*) whose names embody functions based on long use and relationship (and which Linnaeus decided to keep), all plants on Earth now possess these kinds of invented labels that, when used, indicate nothing about their nature. Conventional scientific plant naming creates and sustains the illusion that plants such as osha exist in isolation from the animals, plants, insects, people, and landscapes among whom and in which they grow, that no connections exist between them and anything else. Like all language, botanical language shapes how the world is perceived and the unexamined assumptions that are embedded within it are reinforced the more it is used. Gregory

Bateson expressed concern about how the hidden perspectives in such languaging affects children (and the adults they become) when they are taught it.

> There is a parallel confusion in the teaching of language that has never been straightened out. Professional linguists nowadays may know what's what, but children in school are still taught nonsense. They are told that a "noun" is the "name of a person, place or thing," that a "verb" is "an action word," and so on. That is, they are taught at a tender age that the way to define something is by what it supposedly *is* in itself, not by its relation to other things.[5]

In a like manner Buckminster Fuller took issue with phrases such as "sunset" and "sunrise" noting that the sun neither rises nor sets and that such usage creates a kind of insanity in people by divorcing them from the world in which they live and inculcating a "picture" of the workings of the Universe that is not accurate.

Ecologists have begun to take issue with Linnaeus's system of naming for the same kinds of reasons, insisting that for ecology to succeed "classical taxonomy will have to give way to functional classifications"[6] that essentially would group "together those plants with similar ecological properties, rather than those plants which necessarily look similar, or have similar evolutionary origins."[7] To understand plants and Earth's ecosystems they have to be viewed as *living systems,* not isolated collections of unrelated mechanical bits—an illusion embedded within the language of Western taxonomy. Naming plants instead by their function, by their relationship to their habitat, connects people to that habitat, to the communications and purposes that run through ecosystems. Such naming carries within itself the implicit knowledge of what will happen if a plant is driven to extinction or declines in population. Many older folk taxonomies—often more complex than Western systems—have long recognized that plants play unique and important functions in ecosystems. Their names for them (as with such plants as Elders and Ambrosias) often reflect plant/ecosystem connec-

Intellectual activities have their verbalisms, their confusions and misdirections and these may also accumulate into what are practically diseases. Every science, of course, needs its technical terminology but all have suffered from the verbosity of nomenclatures and, notoriously, botany most of all.

—Sir Patrick Geddes (in Tompkins and Bird, THE SECRET LIFE OF PLANTS)

---

tions and interdepencies and describe more accurately their true nature and functions.

Plants mean nothing in isolation; they are a life-form rooted in and identified by their community, by their relationships to and interactions with all other life on Earth. Individual plants form local neighborhoods and neighborhoods associate together in communities and those group together as ecosystems that interconnect together to form biomes which together form the larger system called Gaia. Ecosystem function determines the plants that grow within them and the nature of plant associations.

Over decades, centuries, and millennia forests move from here to there, wetlands shift location, meadows spread, new species increase in dominance, old ones decrease; everything is in flux. The plants work

---

Wherever [Linnaeus] went the laughing brook died, the glory of the flowers withered, the grace and joy of the meadows was transformed into withered corpses whose crushed and discolored bodies were described in a thousand minute Latin terms. The blooming fields and the storied woods disappeared during a botanical tour into a dusty herbarium, into a dreary catalogue of Greek and Latin labels. It became the hour for the practice of tiresome dialectic, filled with discussions about the number of stamens, the shape of leaves, all of which we learnt only to forget. When the work was over we stood disenchanted and estranged from nature.

—Raoul Francé (in Tompkins and Bird, THE SECRET LIFE OF PLANTS)

with much longer timescales than we do and the cues for emergence of a specific plant species in an ecosystem might only happen once in decades or centuries.

Ironwood and the Sonoran Desert Community

No one living has ever seen an ironwood grow from seed in the wild; the trees can be more than one thousand years old. They often grow in the harshest areas of the southwest Sonoran Desert of North America. The events that allow their seeds to germinate and survive occur decades or centuries apart. But once an ironwood seed germinates, it begins to alter the desert soils around it. It begins creating life and community.

The seedling sends out chemical cues. *Rhizobium* bacteria respond and begin to form nodules on the roots, making nitrogen available for the unique chemistries to come. Fungal symbionts begin to appear and spread a mycelial network just under the soil. A rich community of organisms—one that does not normally exist in the hot and dry desert soils of the Sonoran—begins to settle into the ironwood soil. The process is slow, though; the desert teaches nothing if not patience. In a good year the germinating seed can grow a foot. Not many years are that good.

When the seedling emerges from the soil, it immediately sends out spreading branches covered with leaves. Unlike other trees in the desert, ironwood does not lose its leaves in the hottest parts of the summer; the shade it produces lasts year round. The cooling its canopy creates is constant.[8]

While the canopy forms overhead, the new root system shoots rapidly down through the desert soil, tapping in to the water that runs deep below. Over time, ironwood roots can probe to depths of 165 feet below the desert surface. And like all trees, ironwood brings up significant amounts of water each day and breathes it out through its stomata. At night, when its stomata are closed, the tree releases and stores the hydraulically lifted water and deep-soil minerals just under the surface of the ground.

Smaller plants begin to appear. Continually shaded from the desert sun, cooled by transpired water, and watered daily by hydraulic lift, some 65 species of plants will come to grow under ironwood. Catclaw, elephant tree, ocotillo, creosote bush, wolfberry, desert lavender, ratany, ambrosia, cassia, orange-velvet mallow, saguaro, senita cactus, prickly-pear, climbing milkweed, passionflower, evening primrose, buffel grass, spiny aster, tansy mustard, desert bedstraw, plantain, silver nightshade, and more.[9] Thirty-one of these will grow nowhere else. This emerging plant community connects to the mycelial network and plant chemistries flow throughout the network. Wherever plant roots touch, they can share their chemistries directly.[10] All the plants exude volatile aromatics. Some aromatics call pollinators, others fall in a continual rain over the plant community and to the Earth below. The soil takes them up; the companion plants under ironwood breathe them in. The smaller community plants cover the ground, keeping the soil moisture high. They all release their own unique mixtures of phytochemicals that blend together with ironwood's in maintaining the microclimate and soil community under the tree.

> *By modifying the habitat under its branches, ironwood plays a leading part in creating the ecosystem that it occupies, greatly enhancing the diversity of the Sonoran Desert. . . . The unique environment that ironwood trees create under their canopies is found nowhere else.*
>
> —Tewksbury and Petrovich, "Ironwood as Habitat Modifier Species"

As leaves, bark, and limbs age, they fall to the Earth, forming a layer of decaying matter. Over the centuries, the tree and its community build up a mound of detritus around its trunk and under its canopy, in effect becoming an island or archipelago of life and richness amid the desert—a facilitative nucleus of life. Scores of insects, birds, and animals come to the archipelago. They pollinate, spread seeds, build nests from archipelago plants, dig burrows, mate, aerate the soil, use plant chemistries in their growth, as their medicine, as their food, and contribute, over the years, tons of their own "night soil." Ironwood increases the abundance of life by 88 percent and species richness by 64 percent in any area in which it grows. Plants such as the endangered saguaro cactus can rarely germinate outside the kind of zone that trees such as ironwood create. Ironwood, and similar trees, literally create the ecosystems in which they and other beings live.

The ironwood is powerful, long-lived, and patient. It exerts strong and long-lasting stability in the desert ecosystem. As ecologists Burquez and Quintana observe: "Chance events like sustained droughts or wet periods are not likely to alter the permanence of the community given the homeostatic effects of long-lived ironwood trees."[11] It is not unusual, however, for such stable communities to take a century or longer to form. Many of the plants that grow under ironwood can only do so after other plants have already been there awhile, added their unique chemistries, and altered soil community structure. Thus, the older the tree, the more complex and stable the plant community in which it grows.

This kind of plant relationship or community dynamic is not unusual. It is just easier to see in the Sonoran Desert because of the isolated nature of ironwood archipelagos. The Sonoran Desert has at least three distinct types of archipelagos forming around three different dominant or keystone tree species: ironwood, mesquite, and paloverde.

Elder, Nurse, and Community Plants

Dominant or keystone species such as ironwood are usually large, and quite often they are trees; always they contain more mass than other plants in an ecosystem. Fewer of them (in terms of species) live in the local ecosystem and often scores of subordinate or community plants are associated with them. In many ecosystems the density of plant species growing between the archipelagos is much greater than in the Sonoran and they blend imperceptibly into one another. A hundred acres of meadow might have three primary bush types scattered irregularly within it and fifty to one hundred subordinate plant species in different densities associated with each of them. Even in a forest each tree has scores of plants associated with it. They each form archipelagos; the tree archipelagos in forests are just very close together. Tightly coupled feedback loops exist between all archipelagos in an ecosystem. The intercommunication between them shapes and maintains their ecosystem and its responses to stressors. How ecosystems respond to environmental stressors changes completely with which keystone species are at the centers of their archipelagos.[12]

The formation of archipelagos—the movement of keystone species in ecosystems such as the Sonoran—often takes place over a 400- to 500-year period and is initiated by episodic ecosystem events that are unknown and are not predictable. They are initiated by environmental feedback cues whose nature is unrecognized and unsuspected by the majority of researchers.[13] Quite often, before it can establish itself in a new location, a keystone species must have a plant that goes first and prepares the way. These initial species, usually selected from among the community of plants that grow with keystone plants, are the outriders, the plants whose emergence signals the movement of plant species in mass, and the slow shifting of ecosystems in response to Gaian feedback loops. These plants move first and essentially determine what keystone plants will grow where and when. In a way, they act as "filters" through which keystone species are sifted.[14] This is often done by plants such as the *Artemisias* and *Ambrosias* who, when the soil is ready, send out chemical cues telling the keystone species where and when to send its seed. Though wind, ants, and burrowing animals may sometimes disperse keystone seeds to the new locations, researchers have found that mere wind and animal dispersal patterns cannot explain how the seeds move. The distances are too far, the dispersal patterns too unusual.[15] But by whatever means, the seeds answer the chemical call sent by the nurse plants.

In the Great Basin of Utah and Nevada, sagebrush, an *Artemisia,* establishes the community that piñon pine needs. It nurses the piñon until it is old enough to grow on its own: changing soil chemistry, providing unique chemistries for the emerging seedling, protecting it from the vagaries of life.[16] And in the Mojave Desert approximately three-quarters of young creosote bushes are found under *Ambrosia* shrubs.

Paloverde trees also usually germinate in areas first inhabited by *Ambrosia* species. The *Ambrosia* acts as a nurse plant, raising the paloverde seedling until it is large enough to live on its own. And though the endangered saguaro cactus will establish itself under ironwood it more often prefers paloverde. Thus the saguaro depends on the *Ambrosia* that comes, unseen by modern eyes, sometimes centuries before.

Ambrosia species received their names from the ancient Greeks and Linnaeus, out of convenience, kept them. It still embodies ancient understandings and functional relationships. The name literally means "not mortal" and is named after the food of the gods that, when eaten, confers immortality. Ambrosia species are givers of life. (Commonly, Americans call them "ragweeds.")

Reality is a seamless whole where virtually everything affects virtually everything else. There are, however, various concentrations of interaction or causation, and we have somewhat artificially divided these up into "disciplines." There is a certain amount of overlapping at the edges of many closely related disciplines, and this is good. There is a certain amount of bridging that is done even between more distantly related disciplines, and this is also good. But there is much about the structure of reality that is missed by this artificial classification. There are important connections between information fitted into separate disciplines that are being badly overlooked. These weaken our man-made structure. And there are important gaps in the seams between adjacent disciplines. We have a leaky structure where information that we need to encompass is leaking out. . . . And these inherent weaknesses in our way of dealing with reality by dividing it into self-contained, graspable chunks, become magnified by social interactions within each discipline that tend to draw it into itself and thereby widen the gaps: the tendency of disciplines to develop a jargon and often a dogma and to some extent a clique, all of which make it more difficult to bridge the gaps. This is further magnified by a tendency for research not to push out the borders of the discipline at the edges, nor to establish connections to other disciplines, but rather to superspecialize and plunge ever deeper into the minutiae of the subject until it is impossible to be an expert except by spending all one's effort in the field, with little or nothing left over to become even cursorily familiar with other disciplines.

—Louis Pascal, "What Happens When Science Goes Bad"

Keystone species, once established, call to them not only soil bacteria and mycelia but the plants they have formed close interdependencies with over millennia. As the plants arrive, the keystone's chemistries literally inform and shape their community structure and behaviors.[17] This capacity of keystone species to "teach" their plant communities how to act was widely recognized in indigenous and folk taxonomies.

Elder trees (*Sambucus* spp.), for example, are keystone species in many ecosystems. Among many indigenous and folk peoples it is said that the Elder tree "teaches the plants what to do and how to grow," and that without its presence the local plant community will become confused. This function is contained within its common name, Elder, which comes from the Old English *eldo* meaning "old age." The connotation of the word is that of wisdom and the ability to teach and help the young of the community, to shape their knowledge, behaviors, and relationships to other members of the community. ("Elder" is perhaps a better name for this ecosystem function of plants than either "keystone" or "dominant.") Other indigenous peoples, recognizing the nature and function of keystone species, have said that "the trees are the teachers of the law." Keystone species regulate the broad community dynamics of a plant community (its character), while the smaller community species regulate the flow of life to and through it: pollinators, "critical pests, pathogens, herbivores or mutualists," etc.[18]

Biodiversity and the Health of Ecosystems

The kind of plant succession, plant nursing, and establishment of dominants and subordinants found in the Sonoran Desert occurs in all land ecosystems: desert, wetland, prairie, subalpine, boreal forests, arctic alpine, and tropical.[19] The same kinds of increases in species richness and abundance occur in all of them as well.[20] And the ways that these plant communities assemble *always* produces the most vitality, abundance, diversity, and richest plant growth possible in that ecosystem.

Though community plants (researchers usually call them subordinate or filter plants) grow smaller when in relationship with their keystone species than when alone, and the keystone species larger, together they create *more* biomass than if grown separately, even if supplied with all the water and nutrients they need.[21] As well, they draw down more CO_2 than if they are grown separately, create more extensive root systems, have more dense canopies (thus are more efficient at photosynthesis as a group), more efficiently retain water in themselves and in the soil, and create more complex chemistries. The soil communities underneath them are always more diverse and healthier.[22] Basically, the more diverse an archipelago the more vital and healthy it becomes. The outcomes produced by such diversity are not merely additive, they are exponential: A plant community is far more than the sum of its parts.[23] The addition of every new species that joins an ironwood community contributes to community dynamics in ways that cannot be predicted from knowledge of the individual species alone. Each is synergistic with all the other species already present in the neighborhood.

The industrialized eye, accustomed to suburban lawns and controlled gardens, generally sees such diverse, visually complex plant communities as chaotic. There appears to be no order or control, only wild, random growth. Regardless, plant communities have spent some 500 million years learning their craft; there is a reason for how they are structured. The more visually complex a plant community, the better it can respond to ecosystem demands and stressors. All ecosystems are dynamic over time in their drive to preserve this kind of "wildness." A suburban landscape, not continually forced into an orderly shape, will "relax." It will begin rearranging, reassembling, itself immediately; it will begin to look rather unkempt.

Disease outbreaks and heavily destructive insect infestations are extremely rare in these kinds of lumpy, visually complex, "unkempt" ecosystems. The complex chemistries present in nonmonocropped natural systems automatically limit the emergence of disease and insect epidemics *and* resistance. In healthy plant communities susceptible plants are always located near to those with different chemistries—different medicines. Out-of-control infestations are *always* the result of reducing what appears to be wild chaos in natural ecosystems, of engaging in farming of monocultures, genetic uniformity in plant communities, heavy pesticide or pharmaceutical use, or environmental stress from things such as suburban building or logging. In healthy systems, plant locations shift, the community reassembles in new ways around its keystone species on a regular basis, and the plants never act to kill off all of a "pest" species, they only moderate its population levels.[24]

Many Native American cultures have understood that plants are communal beings, that they grow better together than apart, and that "unkempt" agricultural landscapes produce better results in the long run. Corn, beans, and squash—often called the three sisters—have been grown together in Native communities in the Americas for millennia. Food production over time is always higher than when crops are grown separately and diseases and infestations are less frequent. Indigenous peoples have long known that the plants support each other, keep each other healthy, and that the wild plants around the periphery of the fields help keep them vital.[25] Among the three sisters, corn is the dominant plant, and it is said in many indigenous cultures that "the Law is in the Corn." Cultures that have had long relationship with corn have also said that corn is the source of the Original Instructions about the nature and balance of community—for people and plants.[26] Ecosystems, to be healthy, must be composed of many plants that are working together in such close-knit communal relationships. There is, in such systems, always a diversity of plant species *and* a diversity of functional types.

The larger the number of plants with diverse chemistries that occupy the largest number of ecosystem functional categories, the more vital and healthier the ecosystem. This is because no year is ever the same as a previous year; environmental conditions are always different. Local habitats are always shifting in response to changing conditions. A plant producing major contributions to habitat need one year may not play the same role when environmental factors change. This is why a large range of plants in all local communities is necessary: to give the system maximum response ability.[27] Ironwood archipelagos exert such broad ecosystem stability because, in maturity, they contain a highly integrated plant community that contains maximum response capacity. Like bacteria, they have to have the ability to respond to environmental stressors that they and the Earth have no way to predict. Diversity, unkemptness, and community complexity are all aspects of this capacity to respond. So are exceptionally rare plants, which all mature plant communities, including ironwood, contain.[28]

These rare plant types exist in small quantities in ecosystems and often possess unique or highly potent chemistries. Such uncommon plants represent gene pools of chemistries needed only in tiny quantities or at rare intervals. They represent rare plant pharmaceuticals for conditions that may only be encountered once a century or for conditions where only unusual chemistries are necessary.[29] Similar plants, even though closely related, may not be functional equivalents. They are not, as many researchers insist, merely "extra" or redundant parts. Earth sees needs in decades or centuries; Gaia takes a long view.

The closely related Mesquite- or Grama-grasses *Bouteloua gracilis* and *Bouteloua eriopoda* are a case in point. *B. gracilis* can live a long life, at least four hundred years, though no one knows exactly how long. It grows throughout the Great Plains from Canada to Mexico and as far west as western Texas. *B. eriopoda* rarely lives forty years, and though its range overlaps that of its cousin it grows further west, extending into the southwestern deserts of Mexico, western Texas, New Mexico, and the edges of the Sonoran Desert. Though both grasses grow similarly, in separated "clumps" with spaces between, the two plants affect nitrogen content of the soil, among other things, in significantly different ways. The Plains grass, *B. gracilis,* because of its long life span, concentrates large quantities of nitrogen from the interplant zones beneath it. It covers only 35 percent of the steppe but uses most of the nitrogen from the remaining 65 percent of the soil. Deserts, however, have very little nitrogen available in their soils.[30] In fact, the semiarid grassland steppe that *B. gracilis* occupies has three times as much nitrogen as the desert grassland habitat occupied by *B. eriopoda*. Both species of Grama-grass concentrate nitrogen but the rate is so slow that it takes fifty years to seriously accumulate and deplete interplant zones, a period that exceeds *B. eriopoda*'s life span. If the longer-lived species grew in abundance in desert ecoregions nitrogen would be depleted from the soils and its presence would seriously impair the growth of the other plants in the ecosystem. People who argue that many plant species are redundant (and therefore, not necessary to protect) because there are other, closely related species that can fulfill the same functions are missing the point. Closely related species *always* differ along some axis of variation and there is a reason for this: Environmental needs are different. Otherwise Earth would have simply developed one plant *not* two.[31]

The drive to maintain complex plant types in order to respond to ecosystem demands extends, as with bacteria, to gene structure. Under pressure plant genotypes can go fluid and reassemble in order to enhance flexibility.

> *Received wisdom is sometimes just closeted ignorance . . . ignorance of the wide repertoire of responses found in plants sometimes elicits derision of the term "plant behavior" from the incognoscenti.*
>
> —Jonathan Silvertown, "Plant Phenotypic Plasticity and Non-cognitive Behaviour"

Genetic Fluidity

Plants, like bacteria, can, under environmental pressure, rearrange their genetic structure (the genotype) and produce new physical expressions in themselves or succeeding generations (the phenotype). The Nobel Laureate and corn researcher, Barbara McClintock, was the first to show (though Goethe had understood it much earlier) that plants could reorder their genetic structure under stress. She found that stressed organisms engage in genetic changes not predictable from their genetic makeup. Plants rearrange their genetic makeup, in part, through the use of transposons or "jumping genes." McClintock's biographer, Evelyn Fox Keller, comments: "Where do the instructions [for gene rearrangement] come from? McClintock's answer—that they come from the entire cell, the organism, perhaps even from the environment—is profoundly disturbing to orthodox genetics."[32]

The complex assemblage of plants in ecosystems represents a huge range of genotypes, all kept in a diversity bank. Not only can plant species create and release a wide range of chemistries, or arrange themselves in a multitude of combinations (or neighborhoods) to produce different synergistic chemical combinations, they can also allow their genetic structure to go fluid in response to environmental stressors, producing highly variant offspring in short periods of time. (To

The variation of plant forms, whose unique course I had long been following, now awakened in me more and more the idea that the plant forms round us are not predetermined, but are happily mobile and flexible, enabling them to adapt to the many conditions throughout the world, which influence them, and to be formed and re-formed with them.

—Goethe (in Tompkins and Bird, The Secret Life of Plants)

---

create even more variability, viruses can insert unrelated DNA sequences into plant genetic structure, giving the plant a wider range of responses, or two or more organisms can come together, engaging in symbiogenesis, to create entirely different life-forms, or both.)

Environments can be unpredictable; Earth has no way of knowing what events it will encounter in its long life. By designing in genetic fluidity, all organisms (and the system that is made up of their complex feedback loops) possess maximum responsiveness.[33] The information that plants take in from their environment is used in a tightly coupled feedback loop to adjust genetic structure and phenotype (and the resulting chemistries) with each generation. Plants that are transplanted into stressed environments (or offspring that germinate there) can shift their structure to produce new plant characteristics that can respond to the *exact* nature of environmental demands. Like human immunity to certain diseases, these characteristics can be passed on to offspring and represent the inheritance of acquired characteristics (something insisted to be impossible by Neo-Darwinians). Plants can change branching patterns, nutrient uptake processes, photosynthesis capacity, and chemistry.[34] They can also adjust themselves to the capacities of the animals and insects that live within their community. They literally change themselves, in very short periods of time, to better maintain relationship with specific species.[35] They can also form coevolutionary bonds over short periods of time in response to informational looping in order to adjust ecosystem functioning.[36] Plants can also limit many of their behavioral responses, including genetic fluidity, to make sure that disease or insect resistance to protective chemistries does not arise.

For example, during spruce budworm infestations, spruce forests always contain trees that *do not* produce alterations in terpene chemistry. Researchers examining the trees have found that they *can* increase their production, they simply do not. In other words, these are not "weaker" trees that are simply succumbing to a Darwinian survival-of-the-fittest dynamic, but strongly healthy trees that are intentionally *not* increasing chemistry production. The long-range benefits of this are clear: By not raising antifeedant actions in all the trees, the forest makes sure that resistance does not develop in spruce budworms as it does in crop insects exposed to pesticides.[37] Plant communities literally set aside plants for the insects to consume so as to not force genetic rearrangement and the development of resistance.

Insects such as the spruce budworm are essential parts of plant communities, they are not simply meaningless pests that arose in a vacuum and are trying to wipe out all the spruce trees in a voracious desire to breed and feed. Plants maintain neighborhood, community, and ecosystem health, including insect and animal population densities and health, through their biofeedback mechanisms. This includes providing plant chemistries at the microlevel, for individual members

---

Any given environment usually contains several plants with widely different attributes, showing that there is seldom a unique solution to a given set of environmental challenges. This is consistent with the finding that complex, non-linear, highly linked systems (such as plant metabolism) have multiple stable states.

—R. J. Scholes, et al., "Plant Functional Types in African Savanna Grasslands"

In the absence of a complete understanding, this apparent species redundancy is best regarded as a system adaptation against a highly variable and unpredictable environment. In many cases it is likely to be an essential component of, and in fact may well be used as a measure of, the system's ability to continue functioning when stressed or disturbed.

—B. H. Walker, "Functional Types in Non-equilibrium Ecosystems" of their communities. Animals and insects are not redundant, they are interwoven into plant communities as deeply as plants are in archipelagos. Plants communicate with and call bacteria to themselves, they call mycelia, they call other plants, and they also call insects and animals, joining them to the community they are creating.

Saguaro and senita cacti, for example, often grow within the ironwood archipelago. Both have a unique and specific relationship with their own species of *Drosophila* fly. Both cacti create and release, as a volatile, tiny amounts of a steroidal compound called schottenol that attracts *Drosophila* flies. The flies use the steroid to make their molting hormone, without which they cannot grow to maturity and reproduce; the cactus that calls them is the only source of this substance. To make sure that only one species of fly comes to them each cactus also releases another, different compound simultaneously. Senita exudes minute amounts of an alkaloid, lophocereine, that is repellant to all species of *Drosophila* except for its coevolutionary partner, *D. pachea*. Saguaro, on the other hand, releases the alkaloid carnegeine, which repels all *Drosophila* except *D. nigrospiracula*. Carnegeine and lophocereine possess only minute differences in their chemical structure, yet they are so specific that for every 6,803 larvae on a saguaro cactus only one is not *D. nigrospiracula*. In return for their molting chemistries the flies and fly larvae eat decaying parts of the cacti, keeping them healthy by removing decomposing matter. Because both cacti also emit strong secondary chemistries to keep other insects away, this maintenance function can only be performed by the fly they call.[38] The flies, like the community plants, bacteria, and mycelia, become part of an ecosystem community that can last thousands of years; their coevolutionary relationships with the cacti are millions of years old.

Regulatory interdependencies such as this are the norm throughout the planet's ecosystem. Without these plant chemistries—millions of years in the making—released at specific times, in specific combinations and amounts, no ecosystem could function or remain healthy. Nowhere is this so obvious as in pollination.

Pollination

Without pollinators, most angiosperms, or the flowering plants that emerged 140 million years ago, cannot set seed. They need someone to take pollen from one plant or flower to another. Most angiosperms have developed coevolutionary, mutualistic relationships with different pollinators to accomplish this task. These pollinators, hundreds of thousands of them, are all a deeply interwoven part of plant communities. All the flowering plants, including saguaro and senita cacti, create a sophisticated mix of chemicals that communicate with and maintain their pollinator communities.

Although most people know that honeybees pollinate plants, few are aware that there are hundreds of thousands of plant pollinators, many of whom have coevolved over millions of years with specific plants. Less than 6 percent of them are known to researchers. They range from bats, to mosquitos, to mice, to ants, to opossums, to bees, to monkeys, to beetles, to lizards, to flies, to birds, to butterflies, to flying foxes. There are at least 1,500 bird species, 15,000 wasps, 40,000 bees, 20,000 butterflies and moths, 14,000 flies, 200,000 beetles, 165 bats, and 300 other mammals that pollinate plants. Perhaps 80 percent of all the flowering plants are pollinated by beetles. Forty percent of the angiosperms have a primary, dominant pollinator with a number of other, less regular pollinators. Some plants have only one. The 700 to 900 fig species in the world (including the Indian banyan), for the past 40 million years, have each been pollinated by its own individual kind of fig wasp. Their lives are often mutually interdependent; neither can survive without the other.

The vast majority of pollinators are called to their plants by specific chemical compounds made uniquely for them, which are active in extremely tiny quantities. The fruit fly *Dacus dorsalis,* for example, will respond to as little as 0.01 micrograms (one hundred millionth of a gram) of the pheromone compound methyleugenol produced by *Cassia* plants. The compound is so specific that even minute changes in its chemical makeup will produce little or no response from the fly.[39] Because of this kind of specificity, plants create a wide range of volatile compounds that appeal to pollinators. The smell of rotting meat, fresh feces, decaying fruit, or aromatic floral scents all contain chemical cues that pollinators respond to.

However, most chemical cues, millions of them, have no "smell" at all. They are picked up by receptors that in vertebrates are called vomeronasal organs (VNOs). In mammals, VNOs are located in the nose and their only function is to take in and transport to the brain the complex aromatic chemicals given off by plants and animals in tiny quantities. (Most Neo-Darwinians regarded these as present only in babies, where they labeled them vestigial—no longer useful, former evolutionary stage—organs. In spite of centuries of research on the human body, the recognition that all people have these organs, occurred only in the last decade of the twentieth century.) They are so sensitive in human beings that as little as one ppm of a woman's sweat in distilled water dabbed with a Q-tip on another's upper lip will stimulate her body to begin menstruating in harmony with the first.[40] All pollinators, all land life, possesses VNOs or their equivalent; they are the receptors for the complex chemical communications expressed from plants.

Many plants facilitate this process. They raise the temperature in their flowers (often considerably) to increase the volatilization of their aromatic compounds into the air. They increase aromatic production when pollen is ripe and vary the amount and type of volatiles being produced at different times to attract different pollinators. Since night pollinators cannot use visual cues many plants increase their volatiles in the evening. The different compounds diffuse in the atmosphere at different rates, letting pollinators know how far they are from the source. Using these chemical cues, bees are able to harvest over a range of sixty miles and remember every flowering plant and its location.[41] Plant flowers will continue to release, even increase, aromatics until all their seeds are fertilized or until the flower withers and falls. This often takes a great many pollinator visits. Tobacco capsules, for example, contain 2,500 tiny seeds, which require 2,500 fertilizations (which, in the case of tobacco, must occur within 24 hours) in a space less than 1/16 of an inch wide. As soon as fertilization is complete aromatic production and volatilization ceases.[42]

Because nectar is an essential component of most pollinators' diets (for some it may be the *only* component), nectar constituents are altered depending on when primary pollinators are likely to feed and what kind of pollinator is expected. For example, bumblebees need (and get from their flowers) a much higher amino acid content than honeybees because of their inability to consume large quantities of pollen.

Though often claimed by dieticians or physicians to be "just sugar," the truth is that nectars (and the honeys they sometimes become) contain a variety of different types of sugar compounds. These make up only about 30 percent of the nectar. The rest is a complex mix of more than forty compounds including lipids, amino acids, proteins, and scores of unique secondary chemical compounds. Pollinators use the sugar for energy; how much they get depends on their size. The nectar from a single flower, for instance, can give a mosquito enough energy to fly twenty-five kilometers. But the nectar is also highly nutritious; the plants create and add compounds to meet the nutritional needs of their pollinators. In butterfly-pollinated flowers the nectar contains the ten primary amino acids the butterflies need to remain healthy. To strengthen the nectar's amino acid content even further butterflies often make a pollen "tincture" by soaking pollen grains in the nectar (leaching out more amino acids) before drinking it.[43] Some plants sequester oils in special compartments in their flowers for their pollinators. When visiting, solitary bees break open the vacuoles, harvest the oils—which are a complex mixture of lipids, amino acids, and other secondary compounds—mix it with pollen from a variety of plants, and

---

Contents of Nectar

Amino acids, proteins, lipids, antioxidants, alkaloids, glycosides, thiamin, riboflavin, nicotinic acid, pantothenic acid, pyridoxin, biotin, folic acid, medoinositol, fumaric acid, succinic acid, oxalic acid, citric acid, tartaric acid, a-ketoglutaric acid, gluconic acid, glucuronic acid, allantoin, allantoic acid, dextrin, and other, unidentified compounds.

feed it to their young.[44] The compounds have exceptionally potent impacts on the growth and immune health of the young bees. Many of the secondary plant compounds that pollinators gain from their plant partners are unique chemistries without which they cannot survive and reproduce.

Some orchids create volatile chemicals that are identical to those given off by aphids. Aphid-loving flies and ants respond to the smell and pollinate the plants. A more common approach, however, is making female sexual pheromones that call their bee, wasp, sawfly, and beetle pollinators. An individual orchid, for example, often produces over one hundred different volatile aromatics, uniquely combined, to call its pollinators. Aliphatics and terpenoids such as saturated hydrocarbons, octanal and nonanal aldehydes, 1- and 2-alcohols, esters, geraniol, citronellol, E,E-farnesol, 2-ketones, acetates, farnesol esters, geranial, citronellal, and linalool are intermixed to produce compounds the bees (or other pollinators) need for sexual reproduction.[45]

Orchid (or euglossine) bee pollinators harvest the volatile compounds from the orchid and transfer them to their hind legs, where they store them as compounds essential for mating. During mating the bees use these pheromones to attract other male bees and together they form small swarms that attract the females (who will only respond to swarms). Each different euglossine bee needs a different kind of pheromone, which the orchids provide. Because the pollinators cannot make these mating compounds themselves, they cannot survive without the orchids.[46] The production by flowering plants of unique chemi-

---

There is considerable diversity in the chemical composition of floral scents released by 13 plant species in the Sonoran Desert, all of which are pollinated by hawkmoths. This indicates a complex interaction between a hawkmoth community and a guild of flowers in which the odour signals are only one facet of the information used for location of nectar resources.

—Frietson Galis, "Ecology through the Chemical Looking-glass"

cal substances that their pollinators need for reproduction is not uncommon.

Many moth and butterfly species are drawn to plants such as ambrosia that produce pyrrolizidine alkaloids (PAs), which they collect, store in their bodies, and use as or to synthesize their sexual pheromones. Some male butterflies store the compounds in their wing hair pencils and use them to attract females. During mating they hover above the female and dust them with the alkaloids. Because PAs are toxic when concentrated these plant-specific moths also concentrate them in their bodies to deter predation by birds and other animals. In a similar way Monarch butterflies sequester cardiac glycosides, usually from milkweed species, in such high quantities that birds will not eat them.[47] More than forty moth and butterfly species use unique plant chemistries in this manner. Without the chemical compounds produced by their plants the moths and butterflies could not survive. And without them the plant could not survive—there would be no one to pollinate them. These interactions can be exceptionally complex and involve multiple plants and insects.

[The] transactions between pollen-producing plants and pollen-moving animals make up a significant portion of what biological scientists are now calling biodiversity.

—Gary Paul Nabhan, The Forgotten Pollinators

Tropical passionflower plants produce extrafloral nectararies on their leaf stems that support a diverse group of ants, wasps, and egg parasitoids that protect them from herbivore overfeeding. *Heliconius* butterflies lay their eggs on the plants and act as passionflowers' primary pollinators. The butterfly larvae feed on the passionflower and sequester alkaloidal compounds necessary to protect themselves from birds after their transformation into butterflies. And while the 45 species of *Heliconius* butterflies each specialize in a particular type of passionflower, the passionflowers intentionally limit their own range of growth. There are never more than ten different species growing in any area. And though similar leaf shapes exist among the scores of passionflower species, none of those ten species will have similar leaves. Each passionflower in that area uses different chemical and visual (leaf shape) cues for its own butterfly species. But this interdependency is even more complex; other local plants, *Anguria* species, come into play as well. *Heliconius* collect food pollen from *anguria,* cross-pollinating the plants during their visits. The *anguria* generate exactly enough nitrogen in their pollen to support the egg production of *Heliconius* butterflies. And though the butterflies collect pollen from other sources it is only the nitrogen from *anguria* pollen that is incorporated directly into their eggs. The *anguria* and passionflower plants coordinate the opening of and chemical emission from their flowers so that they are visited in succession.

Like honeybees, *Heliconius* learn their area extremely well, essentially running "traplines" of flowers from whom they collect pollen and nectar. They live communally and the younger butterflies are taught the flower routes by their more experienced elders.[48] Such specificity as this is not unusual.

The female yucca moth's relationship with desert yucca plants is so specific that she actually hand-pollinates the flowers. After gathering pollen the female moth cuts into a yucca flower ovary and dusts each of the immature seeds within it with pollen. On a few of them she deposits an egg before leaving. When the moth larvae hatch, they eat a few seeds as their initial food before leaving the plant; they never eat them all. This same kind of process is also at work in fig wasp/fig pollination. Unique chemical cues call each of the hundreds of types of fig wasps to their fig species; they will go to no other. They gather pollen from the male flowers, then each enters a tiny hole in the fig ovary and hand-pollinates the immature seeds before laying eggs. Up to 70 percent of vertebrate diets in some forests come from such hand-pollinated figs.[49]

This use of plant chemistries by the life in ecosystems is pervasive. These chemistries are used as molting hormones; sexual pheromones; dietary precursors to protective chemicals; essential proteins, amino acids, and lipids; and healing medicines. Plants *always* produce more chemistries than they need merely for their own health; these chemistries are released into plant communities and ecosystems to maintain them. As with so many other aspects of maintaining ecosystems, plants sense when members of their community are ill, and they offer up chemistries to heal them. These chemistries are either sent through mycelial networks to where they are needed, or chemical cues call the ill animal or insect to the plant who needs it.

Indigenous peoples have long known that the plants heal the animals, plants, and other living organisms in their communities. Folk taxonomies link elephants, bees, monkeys, porcupines, bears, elk, deer, mice, and more to the use of plants as medicines. In 1978, for the first time, a scientist excitedly proclaimed his discovery of an animal's use of a plant as medicine. And so it goes.

Plants as Medicines for All Life on Earth

Deeper awareness of the sophisticated complexity of plant chemistries and the inextricably interwoven connections of the plants and their chemistries to the life around them has begun to reveal to contemporary peoples that plant chemistries are used not only for the plants themselves, but are created and released to heal disease throughout the ecosystems in which they grow.

For example, in plant communities, the closely intertwined feedback loops automatically note when any member of the plant community is ill and the mycelial networks just under the surface of the soil transport necessary chemistries to it.[50] Healthy plants connected to the mycelial network increase their production of whatever chemistry is needed and send it to the mycelia for distribution. Trees that are intentionally girdled by scientists (they cut a circle of bark from around the trunk of the tree, which will kill it, so they can see what happens) are supported with nutrients transported through the mycelial network from other plants. They can live for years, while plants that are disconnected from the network die within a year. A wide variety of chemistries have been found to be transported this way, including carbon, phosphorus, sugars, and more. Feedback loops are constantly analyzing the needs of the mychorrizal community, and plants that are ill receive whatever it is they need.[51] Injured plants respond much the same way that people do to analgesics and anesthetics; compounds such as the alkaloids in opium poppies affect plant physiology just as they do ours. Such compounds may be taken up through root systems, transported by mycelia, or breathed in through plant stomata as a gas.

Bee pollinators, most notably honeybees, collect a gummy, resinous substance from trees to make propolis, which they use to coat the interior of their hives to protect it from infection. The resin is collected from scores of trees that exude the compounds onto their surfaces, including aspen, poplar, birch, elm, alder, horse chestnut, willow, pine, and fir. The bees combine the tree resins with nectars, multiple pollens, wax, and the bees' internal enzymes. Propolis is generally about 50 percent tree resins, 10 percent pollen, 30 percent wax, and 10 percent various terpenoid essential oils. It is strongly antibacterial, antiviral, antibiotic, antifungal, anti-inflammatory, antioxidant, and antiseptic, and retains these qualities for many years. Ground-dwelling and solitary bees gather their own combinations of propolis to line their hives, as well. Many of the trees that propolis is collected from, such as willow, birch, aspen, and poplar, exude compounds rich in salicylic acid, which contributes to the strong anti-inflammatory actions of propolis. Bees collect a number of other compounds that they use to keep their hives or nests healthy and functioning. Most are processed from flower nectars and plant exudates. Honeybees, for instance, use flower aromatic volatiles to synthesize pheromones that they use to mark their "traplines" so that other bees from the hive can follow them.

Other insects, such as grasshoppers and certain beetles, use sequestered plant compounds to keep their shells free of bacterial, algal, and fungal infections.[52] Some insects, when infected with tiny mites, specifically seek out medicinal plants to kill the infesting organisms. Many insects collect plant compounds, combine them, and store them in their bodies for later use. Sometimes they use them for protection from predators. The sawfly, for example, gathers terpenes from the pine trees on which it lives, recombines them, and releases them in a spray at predators. Grasshoppers, as well, can combine up to 26 different plant compounds in unique mixtures to produce highly powerful deterrent sprays.[53]

Many birds collect a variety of fresh, strongly medicinal plants and weave them into their nests to prevent and treat pest infestations or boost the immune activity of their young. The birds will separate out just the plants they want from others that look similar, and will gather different ones at different times for different reasons. Starlings regularly pick and add plants that are high in volatile oils or terpenes to their nests, so that the volatile oil emissions constantly bathe them and their young, helping prevent infections.[54] Some raptors, because of the regular presence of decaying meat, regularly harvest strongly antimicrobial plants and weave them into their nests. And many birds harvest and use plants to treat mite infestations in their feathers.[55]

Wild boars in India dig up pigweed (*Boerbavia diffusa*) and eat the roots, which are high in anthelminthic (antiworm) compounds, to control infestations of intestinal worms. Boars in Mexico do the same thing, though they use pomegranate roots that contain the same kinds of compounds.[56]

Dogs have long been known to intentionally eat grass at specific times. They will separate out just the types of grass they want to eat from others that appear similar, altering their choice at different times for different conditions. Quack grass, a favorite of many dogs, regulates the health of the mucous membrane systems of the GI tract, helps kidney function, and possesses antibacterial and antimicrobial actions.[57]

Elephants will knock over a palm tree, step on its trunk, and wait while the foot-sized impression fills with one of the sweetest saps on Earth. Then they will stand, gently rocking in the jungle, until the sap ferments, and then they will drink it and become intoxicated. Elephants also utilize a wide variety of medicinal plants for many conditions, including birthing. Holly Dublin, a researcher with the World Wildlife Fund, spent more than a year tracking a pregnant elephant in East Africa, noting her diet and behavior. The elephant was extremely regular in her habits and food plants until the end of pregnancy, when she traveled 28 miles to a specific tree and completely consumed its leaves and woody trunk. Contractions began and she gave birth to a healthy baby. The plant is commonly used by Kenyan women to facilitate childbirth.[58] Other animals also use plants to assist in birthing.

Pregnant Sifaka females, a type of primate, who usually do not eat plants containing tannins, widely seek out the plants during the weeks prior to birth and consume them. Tannins reduce the incidence of postpartum bleeding during childbirth and promote more rapid healing of vaginal tissues.[59] African Colobus monkeys, chimpanzees, gorillas, and Muriquis (woolly spider) monkeys also seek out a variety of plants to help with certain conditions or heal themselves when ill.

When suffering from pathogenic intestinal parasites chimpanzees will select from a variety of plants depending on the nature of the parasite. For example, they ingest pith of *Vernonia* species to kill and stop the reproductive activity of schistosoma parasites. *Vernonia* contains a variety of potent chemistries, including toxic sesquiterpene lactones and steroid glycosides. The glycosides vernonioside B_I and vernoniol B_I suppress parasite movement and egg laying, reducing their population density. When infected with oesophagostomum parasitic worms, on the other hand, chimpanzees seek out entirely different plants. They pick rough, bristly haired *Aspilia* leaves, which contain a unique compound—thiarubrine A. Thiarubrine A is active against a wide variety of nematodes and intestinal worms that commonly affect chimpanzees. The chimpanzees fold the leaves like accordions and swallow them whole. The thiarubrine A weakens or kills the worms and the unchewed leaves, because of the folding and rough bristles, catch the worms as the leaves move through the GI tract, pulling them loose and out of the system. By not chewing the leaves the chimpanzees ensure they will pass into the small intestine in their whole, folded form and also ensure that the thiarubrine A will not be broken down in the stomach.

Chimpanzees are, as well, exceedingly particular about which part of the plant they choose. They only use the pith of the *Vernonia;* it is lowest in toxic sesquiterpene lactones and highest in the steroidal glycosides. Chimpanzees actively test *Aspilia* plants for activity by holding a leaf in their mouth for extended periods of time before deciding to pick it or go on to another. And they take these medicines early in the morning before they begin to feed so that they can pass through the GI tract unencumbered by food.

For caenorhabditis worms, chimpanzees utilize a variety of ficus (fig) leaves, which they also swallow whole and early in the morning. Only young fig leaves are used; they contain significantly stronger antiworm chemistries. Chimpanzees also use a variety of highly antibiotic plants for other conditions and rub those with antiparasitic actions into their fur to treat skin and fur parasites.

The racoonlike South American coati pick the resin from *Trattinickia aspera* and rub it into their fur to protect against parasites, and capuchin monkeys use at least six plants for the same reason. Capuchins also seek out the leachate formed from rainwater as it runs down the trunks of *Hymenea coubaril* trees and collects at the junctions of major limbs. They use the leachate, which is strongly repellent to insects, as an herbal wash to treat and prevent parasites.[60] Bears, among other things, use osha.

Osha is a plains Indian name for *Ligusticum porterii* and means "Bear Medicine." The plant has been called this for millennia in many languages: the Mexican name for osha, *chuchupate,* is an Aztec term that means the same thing. In early spring, after their long hibernation, one of the first things bears do is to use their long claws to dig osha. Some they eat, the rest is chewed and mixed with saliva into a watery paste, which they spray and rub over their fur. Male bears will dig extra roots, carrying them to their mates, whom they spray and wash as well. Osha root is strongly antihelminthic (active against intestinal worms), and helps clean out the intestinal tract after winter. It is also strongly antibiotic, antiviral, and antiparasitical. Washing fur with osha paste helps clean the body of any lingering winter parasites.[61]

And of course, like bears, people use plants for medicines, for we, like all other life, have long been inextricably interwoven into the fabric of the plant world.

The bear is quick-tempered and fierce in many ways, and yet he pays attention to herbs which no other animal notices at all. The bear digs these for his own use. . . . We consider the bear as chief of all animals in regard to herb medicine, and therefore it is understood that if a man dreams of a bear he will be expert in the use of herbs for curing illness.

—Siyaka (in Densmore, TETON SIOUX MUSIC)

Human beings have used plants for food, clothing, building, and healing as long as we have been. Medicinal herbs have been found in a 60,000-year-old Neanderthal grave, and written records over the past 6,000 years have recorded the regular use of more than 80,000 different plants as medicine. People (like soil, bears, butterflies, and monkeys) have made their medicine by percolating water through plants, eating them whole, soaking them in water for teas, or rubbing them on their skin. They have worked very well for us, and for all life on Earth, for a very long time.

The Danger of Pharmaceuticals in the Ecosystem

The pharmaceuticals created out of the universe-as-machine epistemology present a significant danger to this dance between plant chemistries and the living organisms of Earth. There is a great difference between the two kinds of chemistries. It is not merely a difference of chemical structure but one of *meaning*. Synthetic pharmaceuticals are made primarily for the profit of a few, and secondarily as a way to alleviate symptoms of human bodily conditions defined arbitrarily as disease. Plant chemistries are created out of an intricately interwoven biofeedback communication loop between different elements of the ecosystem in response to changing conditions, in order to regulate those conditions to maintain the homeostasis of the Earth. Both types of chemistries are ubiquitous in the Earth's ecosystems in tiny dosages of ppm, ppb, ppt. Plant chemistries are chemical communication messages; pharmaceuticals are noise. Pharmaceutical production and release, in essence, has embedded within it the meaninglessness that the epistemology of science ascribes to the Universe. But the plant-created chemistries released into the ecosystems of the Earth do possess meaning: each is released to accomplish specific ecosystem purposes. The category error of science as regards pharmaceuticals possesses innate, grave consequences. Human-produced chemicals carry information that is meaningless yet still initiates life web responses. Unfortunately, because we are all expressions of bacterial complexity, they affect the metabolic pathways in all living organisms.

The biochemical steps that generate jasmonic acid (JA) in plants, for example, are extremely similar to those in animals that create prostaglandins—the fatty acids that initiate inflammatory responses to infections in people's bodies.[62] JA-mediated plant responses are initiated where needed by a signaling molecule called systemin. The biosynthetic pathway that produces systemin is itself exceptionally similar to the pathway that produces tumor necrosis factor alpha in mammals.[63] Because of their similar nature both prostaglandin and JA effects are modulated by the common plant compound salicylic acid that is found in aspirin. In short, plant chemistries work throughout the life web because we all share similar chemistries, metabolic pathways, and evolutionary histories. And just because a substance is "natural" does not mean it does not have pollutant effects when concentrated. Natural plant-derived substances are also being put into the Earth's ecosystems in quantities far beyond those normally produced by plants.

Aspirin, noticed first in sewage treatment water in Kansas City in 1976 (along with other substances including caffeine and nicotine), enters waste streams in huge quantities throughout the world.[64] About 100 million pounds of aspirin are ingested worldwide each year—Americans buy 30 billion aspirin tablets alone.[65] All of it is excreted into the Earth's ecosystems. Though a great deal of it is removed by waste treatment, the problem is the dose. In low doses, aspirin enhances plant growth, in larger quantities it inhibits it. Amounts in the range of 10 ppm, as an example, increase nitrate reductase activity in corn, but that activity is inhibited when they rise above 1,000 ppm. Plant species such as the oak *Quercus falcata* produce substantial quantities of salicylic acid in their leaves, which is leached or washed out when it rains, inhibiting plant growth under the canopy.[66] Salicylic acid also interferes with the effectiveness of JA, which plays crucial roles in plant health. JA stimulates phytoalexin production in response to invading pathogens, interferes with insect predators' digestion, and hastens seed set in severely damaged plants. Salicylic acid blocks the production of JA, phytoalexin production diminishes, the impact on insect predators lowers, and ability to set seed lowers or can cease. The amount of aspirin being released into the ecosystem is significant enough to affect all these areas of plant chemistry.[67]

The same kinds of dynamics exist for caffeine. Because of the huge production and consumption of coffee and tea worldwide there is significant excretion of caffeine in places where it would not normally be found. Caffeine is strongly inhibitory of seed germination at even minute doses because it interferes with cell division.[68] The amount flowing through waste treatment plants is sufficient to interfere with ecosystem functioning.

And, of course, synthetic pharmaceuticals compound these kinds of problems due to their pervasiveness and tendency to persist rather than break down. (Though herbal medicines present less danger to ecosystems, the production of hundreds of millions of tons of echinacea in extract form and its distribution to ecosystems where it does not normally occur upsets things as well—as does the removal of large numbers of echinacea plants from their natural habitat. Ecosystems were designed for their inhabitants to derive foods and medicines from local plants and then to excrete them back into common soils.)

Under the pressure of pharmaceuticals, genotypes throughout ecosystems—bacteria, insects, viruses, plants, and more—are going fluid and reassembling themselves in order to reestablish system homeostasis. The more intensely human chemicals act as environmental stressors on the system, the greater the pressure on genotypes in response. This is the primary reason it takes more and more chemicals each year to produce the same level of results in agriculture and medicine. Eventually the amount of chemicals needed will surpass energy supply and the system will contract rapidly in order to reestablish homeostasis. The amount of pharmaceuticals being produced (especially when combined with agrochemical production) is enough to interfere with the homeostatic balance of plant neighborhoods, communities, ecosystems, biomes, and Earth itself. In many instances these chemistries combine together with each other, and sometimes with plant chemistries, in synergistic ways that are not predictable and that produce magnified impacts on ecosystems.

By defining the Universe as a collection of parts of a great machine with no interior depth, we emotionally disconnected ourselves from the community of living organisms on Earth; we disconnected our-

We have tried to remove wildness from the context of our daily lives. We have worked to simplify the natural communities around us, hoping to make our environments more manageable, hoping to be more secure. That has not happened, for a simplified environment is ever more prone to what we call "wild fluctuations"—wild, in this sense, meaning uncontrolled or reckless.

But some people remain who associate wildness with wellness. Piman-speaking peoples of the American Southwest use the terms doajig *for "health" and* doajk *for "wildness." Both words are derived from* doa, *"to be alive" or "to be cured," as are* doakam *for "living creatures," and* doajkam *for "wild and untamed beings."*

—Gary Paul Nabhan, Cultures of Habitat

---

selves chemically as well. Morris Fishbein of the AMA was more correct than he knew when he said that modern scientific chemotherapy is about "to wash away the plant and vegetable debris."

The Dangerous Loss of Plant Diversity

The diversity of plant species has been labeled by most scientists as redundant, and unnecessary; the plants themselves have been labeled resources placed here for the use of Man, insentient, and weeds. These care-less attitudes engendered in people by universe-as-machine epistemology have resulted in tremendous reductions in plant diversity throughout the world—in natural systems, medicine, and agriculture. Such loss of plant complexity interrupts healthy ecologies (internal and external) and allows the emergence of disease everywhere it occurs.

The Monarch butterflies that sequester toxic cardiac glycosides in their bodies as larvae, for example, are experiencing serious depredation from birds during overwintering in Mexico. Many of the more toxic milkweeds that the larvae once fed on have decreased in abundance while other, less toxic milkweeds have increased, due to changes in human land use and agriculture. Because the butterfly larvae are feeding on less-toxic plants they build up fewer defensive chemicals in their bodies and birds can then eat them with impunity. This is affecting the intricate butterfly/larvae/milkweed relationship and subsequently the health of the ecosystems in which they occur.[69]

The internal ecology of living organisms also needs continual inputs from wild plant chemistries to remain healthy. Large herbivores, for example, whose rumens contain more than 100 different types of symbiotic bacteria and uncounted numbers of protozoa and fungi, regularly seek out and eat multiple plant combinations depending on their level of health and particular bodily needs. The ingested plants' secondary compounds alter the composition of the communities in the rumen, input different chemistries, and shift levels of health.[70]

But these kinds of internal ecologies are altered significantly if wild plant chemistries are diminished in the diet through such things as monocropped food plants. Such plants contain considerably fewer secondary compounds than wild plants. In consequence, living organisms that eat them as a steady diet are deprived of the normal complex chemistries that their species evolved with over, sometimes, millions of years. For example, pigeons, when fed a diet that in people causes goiter, will develop polyneuritis—an inflammation in multiple nerve pathways. When the pigeons are allowed to return to their normal ecosystem diet the disease disappears. The birds always harbor the microbes that generate the condition, but the chemical diversity of the plants and insects normally eaten by the pigeons does not allow the disease to occur.[71] The internal ecology of animals is maintained by plant chemistries just as soil rhizospheres are.

In people, increases in cancers exactly parallel the decrease of diverse plants as foods and medicines. In 1900, for instance, more than a hundred different types of apples, fifty different types of vegetables, and thirty different types of meat were commonly found in markets depending on the season. Many were wild harvested and their chemistries were much more highly diverse and potent than the foods we buy now. Many of those wild-gathered plants contained the multiple types of cytotoxic, antimutagenic, and cell-division inhibitory compounds regularly ingested in human diets prior to 1900. The ¡Kung bushmen of the Kalahari Desert, as an example, regularly eat more than seventy-five different plants in their diet in one of the harshest ecosystems on Earth; cancer is virtually unknown. (They additionally work less hours than we do, have a high caloric intake by American standards, and spend most of their time in what we call "leisure pursuits.")

Historically, human beings have eaten between 10,000 and 80,000 different food plants in their diets. Some they intentionally grew, most they wild harvested. The combination of plants shifted from season to season, with local plant populations, and in response to the needs of the people. The kinds of diseases we see now are virtually unknown with such diets. Complex plant combinations keep disease conditions in check in people, just as they do in pigeons and in ecosystems.[72] Today Americans regularly eat less than ten plants in their diet, and many eat less than five.

This kind of decrease is present as well in the practice of medicine. In response to the pressure of scientists and physicians, plants in medical practice have been almost completely removed in favor of pharmaceuticals. While some plants are still used as the raw material

---

Scientists have changed our foods. Take the USDA, for example, they have bred out most of the cancer-preventing compounds in soy. So an average primitive soybean will prevent more cancer than a USDA soybean. This is because we Americans tend to go for bland foods and the primitive soybean has a more bitter taste, so the USDA bred out five different chemicals in soy, and bragged about it. They bragged about lowering the phytate content, the bowman-burk inhibitor content, and the protease inhibitors, the very things that prevent cancer. They bragged about breeding out or lowering the estrogenic isoflavones, which is what soy is getting all the press about these days. They bragged about lowering the levels of saponins and phytosterols. Yet, all these have been shown to prevent cancer. . . . And this happens across the board. Food processors and food scientists are making our food less preventative—not only of cancer but also of cardiopathy.

—James Duke, "Herbal Voices Interview with Jim Duke"

for producing pharmaceuticals, virtually no whole plants are used in modern medicine. Pharmaceuticals, on very short timescales, produce better outcomes than plant medicines, just as the short-term use of pesticides produces better crop outcomes. When viewed on long scales, however, the superiority of pharmaceuticals and pesticides vanishes. The pesticide or agrochemical that produces one season of increased crop harvest also initiates ever widening ecosystem perturbations. Attempting to keep production levels the same as that first year necessitates the use of increasing numbers of chemicals, with greater and greater ecosystem perturbations. Over time, these perturbations result in chaos in ecosystems: drought, desertification, intense weather patterns, and trophic cascades, as complex systems collapse into more simple ones from the loss of species. This escalating dynamic occurs in exactly the same fashion in internal human ecologies and in the human communities that use large numbers of pharmaceuticals. Antibiotic resistance is only one manifestation of the ecosystem perturbations that have been initiated.

Bacteria, however, do not develop resistance to plant medicines. Plant medicines, unlike pharmaceuticals, contain thousands of complex compounds that work synergistically; they are so complex that it is very difficult for pathogenic bacteria to develop resistance. Numerous plant medicines have shown activity against *all* the bacteria that have developed resistance to human pharmaceuticals. Plant medicines also have very few side effects. The African herb *Cryptolepsis,* for example, is more effective in treating even nonresistant malarial strains in clinical trials than pharmaceuticals, and possesses none of the side effects that chloroquine produces.[73]

Plants, like pharmaceutical companies, create highly effective compounds that are toxic when isolated. A case in point is the compound artemisinin isolated from *Artemisia annua* and used to treat drug-resistant malaria. It produces a number of unpleasant side effects during treatment. However, when the whole herb is taken the side effects do not exist and treatment is just as effective. The plant has created several compounds *whose only known purpose* is to alleviate exactly the side effects that the isolated constituent produces in living organisms.

Plants are, in essence, ecological medicines. They do not require expensive factories to make them, they do not discharge pollutants into the environment, they have far fewer side effects (internally and externally), they are renewable, and the knowledge of their use is not held in the hands of a few experts but, like plants in ecosystems, is diffused throughout the cultures that use them. They are also very inexpensive. Stomach ulceration can cost up to $25,000 when treated by surgery, or $1,000 a year with Tagamet. It can be treated for as little as $125 with the use of the medicinal plant licorice. Benign prostatic hyperplasia (BPH), which can cost $5,000 to treat with surgery or $657 to treat with the pharmaceutical Proscar, costs $255 to treat with saw palmetto berries.[74] And of course, in communities that work with wild plant medicines, who pick and prepare them themselves, they cost nothing at all.

Scientists who embraced universe-as-machine labeled indigenous and folk understandings of plant medicines useless superstition and set about removing them from the sum total of human knowledge. The pressure of Western medicine on herbalism, on folk and indigenous healing, has initiated the destruction of systems of biognosis-generated healing hundreds of thousands of years old. It is not so different than the destruction of older forms of agriculture and their replacement with modern agribusiness farming technologies. Universe-as-machine has not created a disease-free life in a Garden of Eden filled with bountiful food, but rather has initiated the increasing loss of plant species throughout the world, loss of knowledge of their use, waste that will not go away, diseases more virulent and pervasive than any known before, and ecosystems in disarray.

It has been estimated that as many as 75 to 90 percent of the world's rural people rely on herbal traditional medicine as their primary health care. It appears neither possible nor desirable to replace this herbal medicine with western medicine.

—Ole Hamann (in Akerle, CONSERVATION OF MEDICINAL PLANTS)

Ecosystems such as wetlands, oceans, forests, and deserts are much like the organs in our bodies and, through complex feedback loops just like ours, help maintain organism health. Loss of plant species causes organ deterioration in ecosystems and biomes, just as it does in us (through diminishment of our food variety).[75] While forests are like our lungs, wetlands are much like our livers. (Perhaps it is more correct to say, since we are so much younger, that our lungs and livers are like forests and wetlands. And do not our brains bear an amazing resemblance to the complex, spherical, intricately interconnected, synaptic-junction-filled living system of Earth itself?) While livers cleanse toxins out of our blood, wetlands purify and remove toxins from Earth's waters. Wetland-specific plants such as cattails, willows, and poplars act as filters to process a wide range of toxins such as heavy metals and even dissolved explosives.[76] The plants in wetlands also exert significant control over methane emissions from Earth.

Methane release occurs through the fermentation of organic plant matter by special bacteria—methanogens. If they did not break down the carbon in plant bodies and release it back to the atmosphere as methane, oxygen levels would rapidly build up to life-insupportable levels. This kind of breakdown occurs at high rates in bogs or wetland areas. Wetland plants act much like chimneys that can control the amount of gas flowing through them, funneling methane through their bodies, releasing it from the wetland.[77] If wetlands lose significant numbers of their plants or if significant numbers of wetlands are destroyed, water toxins cannot be processed, methane emissions are affected. Both water and climate deteriorate.

The recognition of plants as medicines for all life on Earth is important. Thinking of them as medicines stimulates thought about what medicinal action(s) they are performing and what will happen if they disappear or decline in population density. Understanding this also engenders the recognition that specific plants can be planted in ecosystems to treat the illnesses they may be experiencing.

Some phytoremediation scientists are taking this route, planting poplars to help wetlands uptake heavy metal pollutants, though like many scientific physicians their bedside manner is often atrocious.

Other people are beginning to understand that if *Achillea millefolium* can stop hemorrhaging and help the healing of wounds that are caused by technology it can, as well, help an oozing erosion gully that is breaking our hearts, or the scarred landscape created by a bulldozer. There is a reason they spring up in some logged forests. Such medicinal understandings can become quite subtle and complex. Combinations of plants, just like those in herbal tinctures, can be planted together to synergistically help damaged ecosystems. Herbalists can play a role in the healing of ecosystems, especially if they begin to understand the similarity between human organ systems and planetary ecosystems. Such folk and indigenous healers, because they tend to see the pattern that connects, because they have a functional sense of aesthetic unity, are important healers for the planet itself.

There *is* a limit, though. Imagine having to perform dialysis on your kidneys daily, then add in a depleted immune system, then add in failing lungs. Eventually every waking moment would have to be devoted to maintaining systems that are supposed to maintain themselves. The initiation of regular fire setting by the Forest Service to control large-scale wildfires is a case in point. Forest systems naturally initiated burns every five to seven years. When we have tried to take over and perform just that one, fairly clear task, the cost is thousands of people-hours, hundreds of millions of dollars, and constant errors that result in wildfires out of control. Even if we could intentionally control the functions of ecosystems (and who of us would want to—it is much more pleasant to lay in the grass than have to give it its medicine all the time), it would never be possible to approach the billion-year complexity and sophistication of Gaia in the generation and application of her medicines. We simply do not have the capacity.

The Elegance of Earth Medicine

For a number of years I lived next to intensive logging. For many reasons I could not move away so I was forced to experience what happens to life, land, and peoples in the midst of large-scale logging. There were many lessons. One: Before too long, the soil and air began to dry out and in a few years uncontrollable fires began. One year in particular the fires burned very hot. Healthy large trees flash-burned from the heat. The soil literally baked.

Early the next spring, wild lettuce, a plant that is common in that area (but never in abundance), covered acres of the burn. It is a plant used much like a mild opium, to stop pain, to soothe, to help sleep. It is also used to help damaged lungs—from dry, irritated coughs to wasting diseases such as tuberculosis. John Evelyn in 1699 commented that "it allays heat, bridles choler, extinguishes thirst, excites appetite, kindly nourishes, and above all represses vapours, conciliates sleep, [and] mitigates pain."[78] Wild lettuce has been used historically to ease the pain of burns. It only grew in such abundance that one year and only in that area. Could anyone of us have prescribed a better medicine?

Plants are being lost all over Earth; some estimates are one species a day. What will Earth and its living communities do for their medicines when they are gone?

Herbelegy | 9

> Almost any edible wild green is richer in vitamins than domestic lettuce. Table lettuce is to wild lettuce what the Protestant's grape juice is to Christ's blood, and plays a similar role. Part of us still knows we need the Wild Redeemer.
>
> —Dale Pendell, *Pharmako/poeia*

I HEAR THEIR NAMES IN MY SLEEP. I HEAR THEM CALLING, "REMEMBER us, remember us, remember . . ."

american ginseng
black cohosh
bloodroot
kava
echinacea
eyebright
goldenseal*

The Geneva 1997 International Union for the Conservation of Nature (IUCN) Species Survival Commission's Red List of Threatened Plants is a six-pound, three-inch-thick book, 8½ by 11 inches, 862 pages long. It lists 33,798 plants of the as-yet-known global flora of 275,000 species. It is heavy, this record of the passing of the Green Nations.

Franklinia alatahama—Georgia
Brachystelma perditum—South Africa
Ceropegia albistepta—Zaire
Cardamine yezoensis—Japan
Scutellaria naxensis—Greece
Lacistema lucidum—Brazil
Oxygonum lobatum—Tanzania

Once we found them wherever we walked. Their songs, heard by deeper ears than the physical, filled the forests. Their smell uplifted us, their medicine healed us, their colors shaped our senses. Where do

*Joanna Macy created the first poem of this type that I know of. Called "Bestiary," it was printed in *Thinking Like a Mountain* (Philadelphia: New Society) in 1988. I always thought one should be offered up for the plants as well. This is my attempt, with respect to Joanna.

they go once they are gone? What holes within us will remain unfilled once they are no more?

helonias root
lady's slipper
lomatium
partridgeberry
peyote
slippery elm
sundew

Lady's slippers—moccasin flowers—are so uncommon now that I have only ever seen one. The flowers are much the size and shape of a baby's slipper. They shape themselves to embrace bumblebees.

Echinopanax horridum—Korea
Gastonia sechellarum—Seychelles
Megalopanax rex—Cuba
Oreopanax dussii—Martinique
Polyscias cissodendron—Australia
Meryta lanceolata—French Polynesia
Aralia soratensis—Peru

What has it been like to grow, a part of the Earth, for 700 million years? How many pollinators have you spoken to in season? How much of your medicine have you sent into the soil and the air? What other humans have you seen in your time? Did any of them look like me?

trillium
true unicorn
venus flytrap
virginia snakeroot
wild yam
chaparro
elephant tree

You survived the great extinctions of the Cretaceous. Will you make it through the Homosapien? When you are disappearing faster than in any geologic time before?

Gomphostemma grandiflora—Vietnam
Acrocephalus sericeus—India
Acrymia ajugiflora—Malaysia
Monardella neglecta—California
Arischrada korolkowii—Kazakhstan
Acinos corsicus—France
Ballota cristata—Turkey

Embrace me redwood, hold me as if it were for the last time. What will I do out here all alone? Upon what will I rest my eyes when you are gone?

blue cohosh
arnica
calamus root
cascara sagrada
gentian
goldthread
lobelia

In the last minute of recorded geologic time we find the visual record of the growing of life over four billion years on Earth. Rethreading the film we begin to run it backward. We watch and see the species removed one by one in evolutionary inversion. Un-Noahs. Unknowing, we check them off the list of life on Earth's ark.

Hirschfeldia rostrata—Yemen
Iberis semperflorens—Italy
Iti lacustris—New Zealand
Xylosma cordatum—Ecuador
Xylosma grossecrenatum—New Caledonia
Xylosma nitida—Jamaica

Two-thirds of the evolutionary ancestors of our food crops are endangered. Viruses and transposons intermixed the wild and domesticated genes throughout the past ten thousand years so that our food plants remained strong. But the genes are going now, they are going now, they are . . .

maidenhair fern
pink root
pipsissewa
stillingia
yerba mansa

Standing in a forest where I used to find you there is only emptiness. When I was ill I would come to you; in prayer I would dig you and make you into medicine. On a long couch at home I lie, my coughing worse.

Lotus kunkelii—Canary Islands
Fosterella alicans—Bolivia
Hymenorchis javanicus—Indonesia

In long cabinets your remains lie, pressed, flattened, a map accompanying them shows where you used to grow. A scientist says he can make you live again when "we" learn more. What about your only pollinator—the tiny bee with the long antennae that I used to see? Will "we" make him again as well? Who are "we" anyway?

Angelica
Osha

I have never been good at good-byes.

The Lost Language of Plants 10

It is not half so important to know as to feel.

—Rachel Carson

The violet growing along the mountain path
grows for no one in particular
But people cannot overlook or forget it.

—Masanobu Fukuoka, *The Natural Way of Farming*

Watching gardeners label their plants
I vow with all beings
to practice the old horticulture
and let plants identify me.

—Robert Aitken, *The Dragon Who Never Sleeps*

The girl hovered in the doorway, making it a central pivot around which her life's choices circled. Half in, half out; uncertain about committing. Hesitant when she entered.

I invited her to sit. She chose the blue chair. She perched tensely on its edge as if ready for flight, her breathing short and rapid, high in her chest. I offered tea; she said no.

She was young, perhaps twenty-eight, the lingering baby fat of adolescence completely gone, the mature woman's bones in her face struggling to find expression. Her cheeks were drawn, the hollows around her eyes shadowed with pain. The eyes themselves, except when they, with jerky movements, took in my face and the room, were fixed and staring. Her clothes were plain; jeans, a workmanlike white shirt with sleeves rolled to the elbow, and worn, well-used running shoes. Her hair was a lifeless brown.

There is sometimes a porcelain whiteness to the skin of Caucasians that is delicate and soft. Life shines through that skin like a candle flame through the delicate onion skin of a lantern. But her skin was pale, as if never touched by the sun, as if it lived in darkness and in such isolation that some essential life had been taken out of it. It seemed an artificial covering, something worn, unalive, unloved, and untouched.

She was in-between. Between the adolescent youth of the late twenties and the mature woman that comes perhaps at thirty-three, with no map to take her into that uncertain, unfamiliar terrain. The end of one life, another not yet begun. Contractions compressing her life, the cervix not yet dilated enough to permit birth. In a barely perceptible monotone she told me her periods were irregular with intense cramping and heavy bleeding. She told me her hands were always cold.

I looked at them. They lay turned inward, fingers slightly curled, on top of her thighs near the knee. They were pallid and slightly waxy like the skin of her face, their character only beginning to form. Like many young hands only partially alive, rarely inhabited. Her fingers, those sensitive touching extensions of herself, possessed nails and cuticles that were chewed and torn. As if she were turning inward and tearing away at herself. Eating herself alive. The muscles of her forearms were taut and filled with tension—barely holding on. And as we talked, occasionally, her hands would tighten spasmodically and, finding nothing to grab hold of or no reason to become a fist, they would relax again. But . . . half open, curled, ready.

I asked her more questions and she told me that she was in the midst of a messy divorce and that she couldn't sleep and while she talked I noticed she sat slightly hunched over so that her chest was hollowed. As if she could make her breasts smaller, call them back inside her, make them disappear. As if her shoulders could curl all the way around and cover her heart with muscle and bone and strength of arms. And I was beginning to breathe shallowly too, and I thought—*Angelica*.

I told her there was a plant I thought she should meet and so, we went for a walk.

I took my staff but the sun was warm and the sky clear—I did not need a coat. We cut across the meadow in front of the house, skirting the driveway with its hard-packed gravel, and began wandering down the other side of the ridge. The slope dropped rapidly and I angled left, taking us into a milder ravine shaped over time by water running off the ridge. The ravine had silted up: There was a grassy, slightly sloping flood pan, with occasional wild rose and spreading prickly thickets of short, matted red root. After thirty yards or so the ravine spread out into a broad open meadow with an aspen grove at its base. The descent was gentle. The meadow's lower edge was slightly marshy and here the milkweed grew, butterflies thick among them. Scores of scattered plants were taking advantage of the water thrown up near the surface by the underground aquifer (fed by the ridge and ravines above) where it ran into the single, massive root of the aspen grove before us.

The air of the grove was thick with aspen's peculiar resinous smell and the hum of bees busy harvesting the sap to line and protect their hives. The small, heart-shaped, slightly serrated leaves, even on trees deep within the grove, moved in the slow breeze. They were shaking, almost shivering, in unison as if they were dancing with some secret joy known only to themselves. I could feel my spirits lift, my feet dance/walking to their secret joy, as I moved among them. The young woman breathed slightly deeper; the gladness of aspens reaching even into her difficult interior world.

Aspen roots are some of the oldest living beings on Earth but the trees they give birth to are young. They rarely grow more than a foot to a foot and a half thick before dying back. This grove was a small one, about forty feet in diameter, and as we moved down the hill the bright dancing light of the aspens gave way to the deeper and darker pine and fir forest that lay below.

The firs and pines were older, perhaps two or three hundred years of age. The forest thickened as we walked on, the sun's dappled golds only occasionally slanting down through some opening in the overstory. It was darker, more peaceful, silent, slower. We began to breathe more deeply, taking in the sun-quickened musky scents of the pine and fir. As we neared the small year-round stream that wove hidden among the rocks and roots of the ridges the slope began to bottom out. Cow parsnip began to appear, big elephant-eared leaves spreading out from strong muscular stalks. Then, a profusion of plants: violet and figwort, dogbane and osha, and horsetail and poleo mint along the shallows, amid the rocky sand of the stream.

The north-facing slope on the other side was thick and old and deeply green. The firs were tufted with short, gray-green growths of the lichen some call usnea. A few were completely covered, shaggy, shambling green beings wakened from hibernation by our passing, thirty feet tall. There was a great rocky outcropping hidden among them, its sides covered in lichen and shadows, its base a tangle of wild roses and mountain mahogany. And there was a game trail at our feet, just this side of the stream. We turned left, moving along it as it followed the twisting and turning of the water. Occasionally we would happen on huge rotted stumps torn open for grubs by the bear that lived on top of the mountain. Scattered clusters of false Solomon's seal, dug from the base of the stumps where bear had left them, were not yet wilted. Their roots long, pale, knobby, and succulent.

The game trail was only a few inches wide, its soil black and rich, making the plants along its length emerald green in comparison. We sank down slightly in its surface tension as we walked, and then, ahead of us, I caught sight of Angelica.

She was growing right at the water's edge, three or four feet off the trail. Majestic and stately, vitally alive and deeply feminine, six feet tall, perfectly balanced between heaven and Earth. I was intently watching the young woman, who was a step or two ahead of me and slightly to my right. Ducking under a leaning lodgepole pine, she caught a glimpse of Angelica out of the corner of her eye. She stopped and turned toward her.

Something inside her shook itself, visibly awakened, and took breath. Something coming alive, breathing for the first time. She began moving, feet walking of their own accord, an invisible force pulling them together. Her hands fluttered over Angelica's leaves, barely touching, as if they were the body of a lover. She turned her head toward me, eyes questioning.

"What is it?"

"Angelica," I said.

She turned back, her hands still fluttering, touching. Paused. "It's hollow. . . . Isn't it? . . . Just like me."

She looked, to see my face.

"Yes," I replied. "Yes, it is."

Our eyes met, hers expectant, almost-hoping.

"Ask Angelica to come into that hollow space inside you," I said. There was a pause and I watched while she did that thing.

She closed her eyes and her breathing slowed. Then, of a sudden, she took a deep breath. Her chest filled out. She straightened and stood taller, more balanced, erect. Her skin softened, the white color flushing rose. The tension in her forearms relaxed, softened, released. She opened her eyes, they were moist and soft focused.

"Oh . . . Oh . . . I see," she said, her voice filled with wonder.

"Now walk awhile," I told her. And she did, moving slowly along the stream. Her body continuing to straighten, filling out, becoming more mature. I watched as Angelica taught her how to become whole, a mature woman.

There are holes inside all of us. Holes that can only be filled by certain plants. Empty space needing tree or stone or bear. Emptiness that can only be filled by some of the other life of this Earth. Other life with which we have evolved through a million years of coevolution.

Without filling them we live a half-life, never becoming fully human, never being healed or whole or completely who we are. Never becoming completely sane.

After a while she stopped walking and we talked. Then we found another angelica plant further along the stream. And saying a prayer and asking for help she dug the plant—for it is the root that is most often used for medicine.

The root is delicate and somewhat small for such a tall plant. There is a central bulbous taproot, something in shape and color like a parsnip but smaller, with a whorl of smaller rootlets spreading out to hold her stable and upright. The color is tannish, not as white on the exterior as a parsnip. The interior is cream and has a sublime smell, indescribably angelica, aphrodisiacal. Human catnip.

She kept the root close to her, rubbing her fingers along its sides, periodically lifting it to her nose as we made the long walk back to the house. For a month she took a tincture of the root, which tastes much like it smells, and her menstrual periods normalized, the painful cramping slowed and stopped.

Angelica in one form or another has been used for thousands of years to help normalize female reproductive problems, to help with severe cramping or menstrual irregularities. People have used it to help

---

To what does the soul turn that has no therapists to visit? It takes its troubles to the trees, to the riverbank, to an animal companion, on an aimless walk through the city streets, a long watch of the night sky. Just stare out the window or boil water for a cup of tea. We breathe, expand, and let go, and something comes in from somewhere.

—James Hillman, The Soul's Code

The body cannot be whole alone. Persons cannot be whole alone. . . . To try and heal the body alone is to collaborate in the destruction of the body. Healing is impossible in loneliness.

—Wendell Berry, The Unsettling of America

breathing, to alleviate hysterical asthma, as a relaxant, and to facilitate journeying in sacred realms. We humans have known Angelica as long as we have known. Plant medicines have always called us in this way.

Plants and People

From the beginning we have been surrounded by the presence of plants; it is not only upon the soil, other plants, or the insects that the aromatics of the plant world fall. They fell upon our species as well; all during our long growth we breathed them in. The *Artemisias* nursed more than plants; they were outriders for more than trees.

In Africa, southwestern Asia, Greece, Iran, Syria, Spain, and Italy humans lived amid steppes of *Artemisia*. As Delores LaChappelle remarks: "The persistent odor of [sagebrush] accompanied humans as slowly, over generations, they moved further north and into the Paleolithic cave areas of Spain and France. Then, as the climate changed, the persistent odor of the *[Artemisia]* steppe moved further north into the areas where humans later learned to grow cereal grains. Throughout all this period of human development" our species was immersed in the volatile aromatics of the *Artemisias*.[1] They preceded us wherever we went; they have called us to them wherever they grew.

Plant chemistries, unlike pharmaceuticals, are released into the world for a reason. Each chemistry a word imbued with import, all together a language that possesses its own grammar and syntax. Its own underlying epistemology. Scientists study the vowels and consonants and how they are put together to form plant words, but they do not study their meaning, nor the intelligence or intent that gives rise to that meaning. Too often they insist there is no meaning, for, as anyone can see, plants have no brain. Still, each complex assemblage of expressed plant chemistries is a sentence of communication, carrying specific messages, imbued with *meaning*. And the world receives those meanings and responds.

Plant communications are like stones in water. The ripples they create move throughout ecosystems; they wash up against us. That we take plant words in through our nose or our skin or our eyes or our tongue instead of our ears does not make their language less subtle, or sophisticated, or less filled with meaning. As the soul of a human being can never be understood from its chemistry or grammar, so cannot plant purpose, intelligence, or soul. Plants are much more than the sum of their parts. And they have been talking to us a long time.

Do you think it possible to dissect a human being,
render it down into its constituent parts,
feed them into a machine which measures such things
and determine from that
its ability to paint or create great music?
No?
Then why do you think
that once you have done this with my body
you know anything about me?

Human beings, long embedded within their environment, have always been sensitive to the meanings that surrounded them. Those contained within plant communications, as with all communications, generate *feelings* in us in response. We know the touch of the world upon us, that we have been caressed by meaning, even though we might not be able to consciously say just what that meaning is. A door opens inside, our unconscious gathers it in, and at night we dream and it is woven into the fabric of our lives. We have always been surrounded by such meaning-imbued language; later we created our own. Our language also travels through the air, though it is vibrating waves of sound. (Did you think we made all this up out of our bulging forebrains alone?) We have always lived, surrounded by original language.

The plants release Earth's subtle chemistries through their intertwined, interdependent synaptic feedback loops faster and more complexly than researchers, or anyone, can write them down. Their meanings pile one on top of another until the linear mind is overwhelmed. They shift the fabric of our world, touch us with meanings, in ways too complex for our conscious mind to grasp. Nevertheless, we can grasp them. They come in gestalts to the attentive mind and caring heart, concise reflections of plant and Earth interrelationships, in knowledge whole and complete: "The trees are the teachers of the law." It is not necessary to have a degree in English to understand the meaning of words. Training in chemistry is irrelevant to understanding the meaning conveyed by plants. A four-year-old does not worry about it; she sits under a tree talking with flowers. For thousands of generations, human beings in all cultures on Earth have known that plants (and *all* of nature) express meaning and that there is intent behind it. They have been listening to and accumulating those meanings; they have been building a biognosis-generated oral library of knowledge of the world and humanity's place within it.

Among the Iroquois, it has been said that if a person becomes ill and needs a plant for healing the plant will stand up and begin calling, helping the person who is ill to find it. To a *Heliconius* butterfly, a honeybee, a moth, or the woman with whom I walked in the woods, this would not be strange; they would understand. (In a like manner that woman would also understand what the Seminole meant when they said that the body of the plant helps the body of the sick person but it is the spirit of the plant that cures the disease.)

Researchers have commented that the plants, able to respond instantaneously to ecosystem changes and inputs with shifts in their chemistry, can begin immediately to produce new compounds and combinations of compounds at need. And when researcher Cleve Backster connected a lie detector to a plant, he was astonished that it could tell from his thoughts what his intentions were when he was going to burn it or tear its leaf.[2] To a *Drosophila* fly or a hawkmoth this would not seem unusual or momentous. To the Winnebago and the

---

Scientists are incapable even of knowing that they are in no position to understand the soul of a flower in the meadow. In the belief they are exploring the root of life, geneticists extract and synthesize the genes present in the cells of living things. But nature's soul does not lie hidden within DNA.

—Masanobu Fukuoka, THE ROAD BACK TO NATURE

Iroquois this would not be strange; they would understand. For it has been said among the Winnebago that when gathering plants as medicine, if you tell them what you need them to do and ask them to put forth their strength on your behalf they will do so. And among the Iroquois, it is said that when you find the plant you are looking for you should pray to it for help. It will tell the other plants what you need and when you pick them their medicine will be strong and powerful.

Many scientists have remarked with surprise that Luther Burbank, George Washington Carver, and even the Nobel laureate Barbara McClintock all have said that it was the plants who told them what to do, who revealed their mysteries to them. The only requirement, they commented, was that they had to care for them, to treat them with respect, to have a feeling for the organism. This would not be strange to the Winnebago, among whom it has been said that people must treat the plants like human beings, make proper offerings, and treat them with respect if they wish their help. Nor would this be strange to a four year old sitting and talking with flowers.

Scientists have discovered that plant species may possess widely different chemistries depending on the time of day, week, or month they are picked.[3] And though the physicians laughed at them, the Appalachian folk healers would have understood and been unsurprised. For among them it was common knowledge that this plant must only be picked in the morning before the dew is off the leaves, or that one only by the light of the full moon.

The plants have long been our teachers and healers. The Cherokee and Creek understood this long ago. It was said among them that the plants took pity on the suffering of their offspring, the human beings, and that each plant offered up a remedy to heal one of the diseases of humankind. There is deep wisdom in this. Understanding ourselves to be offspring, the children, of the plants, naturally engenders a familial bond. It shifts the focus of our relationship from one of plants as resources to them being senior, caring members of the same family. More than that, the power lies with the plants, not with us. *We* are *their* children; they are not our property. As with all children, when we hurt, the nature of their relationship to us leads them to want to help us.

When the ancient Greeks named certain plants *Ambrosias*—givers of life—this is something they understood.

The ancient Greek and Roman, folk and indigenous plant names and plant usages that developed over millennia came out of this perspective. The use of plants as medicines connects us as people to those ancient traditions, to the environment of which we are a part and from which we came, to the meanings that the plants generate; for a million years human beings have been healed by them. Wendell Berry has captured the knowledge of this succinctly:

> Herbalism is based on relationship—relationship between plant and human, plant and planet, human and planet. Using herbs in the healing process means taking part in an ecological cycle. This offers us the opportunity consciously to be present in the living, vital world of which we are part; to invite wholeness and our world into our lives through awareness of the remedies being used. The herbs can link us into the broader context of planetary wholeness, so that whilst they are doing their physiological/medical job, we can do ours and build an awareness of the links and mutual relationships.[4]

By having disconnected ourselves emotionally from the Earth and plants we have lost our understanding of these links and mutual relationships. We have facilitated the loss of plant species, the loss of health in ecosystems and our bodies, and the loss of the sense of who we ourselves, are.

The Importance of Restoring Biophilia

The loss of connection to plants, to the land, to Earth, leaves the holes with which we are naturally born unfilled. No matter how much Ritalin or Prozac is poured into those holes, synthetic pharmaceuticals can never fill them; merely human approaches can never heal them. Pathologies come from the empty holes that are unfilled, from lack of contact and communication with the wild. The holes within us possess particular shapes—that of stone or tree or bear. It is not only plants that are our teachers and healers; not only plants that are among our community of life; not only plants that have a language we have long known.

Without deep connection to the land our healers remain anthropocentric—human centered—in their approaches, their theories of human health generated in isolation from the environment with which we evolved. They contain the same category error that all reductionistic sciences contain. The solution is reconnection to the natural world and the living intelligence of land.

Many people believe we should first establish this reconnection in the young. But I think that the best hope for restoring biognosis is with the grown—those in whom the impulse for biophilia has been stunted, those in whom the interior wound is deep, those in whom the need is greatest. Though children express biophilia most naturally and it awak-

---

The "facilitating environment" so necessary [to psychological healing] is the actual *environment, were it not so neglected and therefore feared. By leaving the actual world out of its main theoretical constructs, psychological theory imagines that world out there as objective, cold, indifferent, even hostile . . . We are back to a world of nature as conceived by Descartes four centuries ago, a bare* res extensa, *an extended field of matter void of soul, inhospitable, mechanical, even demonic.*

—James Hillman, THE SOUL'S CODE

I wish that the men now serving the great polished wheels, and works in iron and steel and brass, could somehow be spared an hour to sit under this ancient oak in Thardover South Wood, and come to know from the actual touch of its rugged bark that the past is living now, that Time is no older, that Nature still exists as full as ever, . . . That they might gather to themselves some of the leaves—mental and spiritual leaves—of the ancient forest, feeling nearer to the truth and soul, as it were, that lives on it. They would feel as if they had got back to their original existence, and had become themselves.

—Richard Jefferies (in Kaplin, TONGUES IN TREES)

ens most easily in them, industrial society has a deep and vested interest in its dominant epistemological perspective. It will not look kindly on any effort to alter it in those future employees who are being trained to carry it on. Nor is it likely to allow the schools to be changed except in minor, cursory ways. And those ways will most likely still embody the reductionistic paradigm that is at the heart of American education. Frankly, as well-meaning as many educators are, unless they first free the experience of biophilia within themselves, learn to experience it, let it deepen into biognosis, and begin healing the damage that such repression has caused within them, they are incapable of teaching it to children. And they are likely to do more harm than good in the trying.[5]

The historical teaching of children by elders that Nabhan speaks of

---

Even when schoolyard landscapes invest heavily in drip-irrigated, low-maintenance landscaping, the composition of their shrubbery remains cut off from qualities of the nearest natural habitats. In fact, such overly manicured, formal designs are not habitats at all, in the sense that other life-forms—wildflower, swallow, hawk moth, gecko—might settle down in them. Nor are they habitats for children, who are hardly stimulated by neat rows of lollipop-looking trees. . . . [Can any] "re-creation" occur when children are kept out of touch with the rest of creation?

—Gary Paul Nabhan, Cultures of Habitat

I remember the boredom of analyzing sentences and the boredom later, at Cambridge, of learning comparative anatomy. Both subjects, as taught, were tortuously unreal. We could *have been told something about the pattern which connects: that all communication necessitates context, that without context, there is no meaning, and that contexts confer meaning because there is classification of contexts. The teacher could have argued that growth and differentiation must be controlled by communication, [that] the shapes of animals and plants are transforms of messages. [But he did not.]*

—Gregory Bateson, Mind and Nature

as crucially important in awakening care of the natural world is not the kind of teaching that occurs or can occur in school systems. There is, along with the *information* that elders convey, the direct transmission of an experience from the body of the elder to the body of the child, from one generation to another. Elders, whether biologically related or not, demonstrate in their very attitudes and physical movements, most often without speaking of it, their intimate personal connection with the life web of Earth. Something is entering and leaving their bodies; they are exchanging their soul essence with the world around them. (For not only do communications come to us from the world, they go from us *to* the world as well.) When elders share information about the natural world with children, about the biota that they know, the information comes out of this central and essential experience. When they share they do not talk about unrelated parts in a void of meaning but express revelations about kin in a context of interconnected relationship. There is a pervasive *feeling* in the room where it is occurring that is based in a specific way of seeing and experiencing the world. *It cannot be mimicked.* It must be genuine, alive and real, for it to be effective. For the children are not learning something dead, mere information. They are learning, as John Seed calls it, to reinhabit their interbeing with the world.

If simple information were enough to stimulate the experience, a book would do as well. But books do not and cannot do as well. Elders have a central, experiential reference point—the meaning-filled livingness of Earth and the biota—that they are embodying as they teach. If the teacher is simply some university-trained adult taking the "kids" on a field trip they are more than likely going to denigrate, however unintentionally, the child's natural experience of the living intelligence of nature. Adult, licensed teachers are embedded in a human-centric, we're-the-most-intelligent, universe-as-machine hierarchy of values that is naturally going to conflict with or denigrate the living reality found by children in nature. This denigration, even if it's only passively embodied in the teacher's manner, intonations, gestures, and choices of language, directly interferes with the emergence of biophilia and the development of biognosis in children. Children, like all of us, can pick up, with the mildest of cues, the most subtle beliefs people possess. Their unconscious, like ours, is accustomed to gathering meaning into itself and sifting for understanding. Any disconnection in a teacher will be picked up and internalized. It cannot be circumvented by rote learning, the teacher must embody the experience itself so that the child can observe it in action as the teacher teaches. And the child must have regular access to wild places where she can observe plants and animals doing what they naturally do. When the child then returns to the elder or teacher, the two of them explore, in a shared context, the expressed meanings, the living soulful intelligence they have found and come to know. And elders are not merely older people—adults. Elder status does not simply occur with aging. Elders are those with extensive experience of biophilia and a large biognosis-generated knowledge base. Elders, like elder trees, teach their community "what to do and how to grow" so that they will not become confused. It is the adults who must, in a strange reversal, learn biophilia first.

I think that many adults, driven by something they do not understand, are being called to fill those holes in themselves. They are beginning to "come to their senses." I meet them at herbal conferences, in tea rooms, and in bookstores. And though they may think they are crazy, they sometimes tell me (often in whispers) that once they were sick and a plant healed them and at that moment something strange began to happen to them. Such a thing as never happened before. This healing from a plant reawakens in the human heart the capacity for deep feeling and connection with the natural world. It opens up an ancient world to those who are healed, a world they did not know existed. It stirs dying embers to flame.

Hundreds of thousands of people have trained in herbal medicine in the United States over the past thirty years. It is a profession they are not legally able to practice because of the pharmaceutical/physician alliance that outlawed the practice of herbal medicine in the United States in the early 1900s. (A small number of them do practice in a few states because they are licensed in another healing profession such as naturopathy, midwifery, or acupuncture—professions that were themselves illegal until just a few years ago. But they cannot practice as herbalists.) I have asked hundreds of them over the years why they have spent so much money and time training in a profession that they cannot practice. And though a few say things like, "I have always been interested in plants," or "A plant healed me once and I wanted to know more," the majority say something like, "I don't know, it just seemed I should. I felt I was supposed to." I have always thought this strange in our society, that so many people should spend so much time and money based only on a feeling they could not explain. And then I thought, maybe our genetic structure is going fluid, too, that maybe the plants are calling us back to an understanding of things we are going to need before too long.

For those of us who were taught, and came to accept, that the Universe is a machine, the journey back to wild water is a long one. We find our way one step at a time.

It is only when we are aware of the earth and of the earth as poetry that we truly live. Ages and people which sever the earth from the poetic spirit, or do not care, or stop their ears with knowledge as with dust, find their veins grown hollow and their hearts an emptiness echoing to questioning. For the earth is ever more than the earth, more than the upper and the lower field, the tree and the hill. . . . It is this earth which is the true inheritance of man, his link with his human past, the source of his religion, ritual and song, the kingdom without whose splendor he lapses from his mysterious estate of man to a baser world which is without the other virtue and the other integrity of the animal. True humanity is no inherent right but an achievement; and only through the earth may we be as one with all who have been and all who are yet to be, sharers and partakers of the mystery of living, reaching to the full of human peace and the full of human joy.

—Henry Beston, HERBS AND THE EARTH

Techniques for Restoring Biophilia

The restoration of our capacity for biophilia begins with restoring, and supporting, our capacity for feeling. And not just feeling in the grossest sense—feelings of anger or sadness or joy or fear—but the subtle feelings it is possible for us to perceive, if we desire to, in everything around us.

We were born with a sophisticated capacity for detecting emotional nuances in the world around us. We feel them every time we encounter the messages embedded within the natural world. Restoring biophilia means exploring these nuances. It means "coming to our senses," especially the sense of feeling—of touch—of being touched by the world. The shadings of emotional color that we can sense come from the touch of the world upon us. And these shadings exist in a thousand colors and tones. It is only through exploring the territory of these feelings that what they mean can be understood. It is not an academic or rhetorical process. It has nothing to do with theory. Feeling comes first, thinking second; thinking in service of feeling. The experience cannot be written down nor found in books. It can only be developed by opening up to the sophisticated capacity for feelings that we possess, by allowing ourselves to be touched by the livingness in the world, and exploring the meanings we encounter. This reconnects us to everything around us—to everything that generates those feelings. It reweaves us into the fabric of life.

What follows are a series of exercises that I have used and taught to people for the past twenty years. They may seem dumb or foolish or scary or stupid, even hard and difficult or deeply emotional. They are all of those things. Doing them often and for several years is helpful. So is writing your experiences in a journal.

Exercise 1

Take a day or an afternoon and go to a part of your town that you like. Choose a part of town that is normally fun to you, that you feel happy visiting. You are just going to be walking and visiting stores.

Begin by walking in the area that you enjoy most. Let yourself sink into the feeling of the place; let yourself relax.

Now. Look around you and pick the store that you feel drawn to the most. Go stand in front of it. What feelings do you have? Let yourself explore them; allow yourself to not be in a hurry. Allow any feelings that arise to emerge; notice what they are.

In the beginning this may be confusing. The multisensory nature of human perception and feeling is so commonly repressed that it is often confusing, or scary, or awkward when opening up to it again. Still, allow yourself to notice whatever you feel—don't make any judgments about it. Write everything down.

Pay attention to the door(s). To the windows, to what is in the windows. To the sign or signs. To the walk in front of the store. To any plants or trees that may be growing there. How does each part feel to you? Do some parts feel better than others? Can you tell why? Overall—what is the primary feeling the store communicates to you? Is it prosperous? Comforting? Happy? Somber? Melancholy? Spend as much time as you need to feel like you have explored every aspect of the store with your feelings and come to a conclusion about it.

Now. Look around the street. Pick another store, but this time pick one that, to your immediate emotions, feels significantly different than the first one. Go to it and repeat the exercises.

Compare the two stores. What different kinds of feelings did they

---

When we are really awake to the life of our senses—when we are really watching with our animal eyes and listening with our animal ears—we discover that nothing in the world around us is directly experienced as a passive or inanimate object. Each thing, each entity meets our gaze with its own secrets and, if we lend it our attention, we are drawn into a dynamic interaction wherein we are taught and sometimes transformed by this other being.

—David Abram, "Trust Your Senses"

generate? Can you tell why? Can you put them into words? (This may take some practice.)

Now go to a third store and repeat the exercise and compare it to the two you explored before.

All of us unconsciously choose to go to stores or restaurants that meet emotional desires we have, that we feel most comfortable in, even though many other stores may sell the same things. This exercise is a process of beginning to consciously perceive and identify the embedded communications that come from the world around you and are felt in subtle emotions.

The businesses that people create embody the world perspective, the underlying epistemologies, that their owners possess. The businesses, like the barn in John Gardner's exercise, convey to customers specific *meanings* through the feelings the customers experience, though they may not normally be able to say what those feelings are. It is possible, after much practice, to identify these feelings, the nature of the organizational structure of a business, its level of psychological health, its impact on the public, its level of financial health, and so on.

Exercise 2

Now go to a coffee house that you like—one with a bookstore is good for this exercise—a place you can linger for a while and have some coffee or tea. Choose something you especially like. Choose a table that has a good view of the room and perhaps the people entering the shop. Let your eye find whichever person you are drawn to most naturally. Observe them. Since you will be looking with some intensity you will have to be clever to not make them nervous or wonder what you are doing or why. This works best if you can observe them unobserved.

What kinds of feelings do you get from this person? Happy? Sad? Nervous? Empty? Masculine? Feminine? Strong? Weak? Poverty? Comfort? Assurance? Indulgent wealth? Indulgent emotion? What thoughts come to you when you look at their face? Let yourself examine their face. How does their chin feel to you? Their nose? Their ears? Their eyes? What is communicated from their eyes? Faces are extraordinarily faithful to the internal world of a person, no matter how schooled someone is at "keeping face." Each part of their face, through the feelings you feel, will tell you something about that person's internal world.

Now. Look at their hands. Do their hands seem alive and aware or asleep and unlived in? Are their hands strong or weak, happy or sad? Businesslike or filled with feeling? How old do you think this person is emotionally? Just let a number come. (Have you known other people who seem that age? Are their hands similar?)

How are their clothes? What do these communicate? Their shoes? Is the person comfortable in their clothes? Are they comfortable in their skin? Do their clothes match the feeling you have from looking at their face?

Do this with as many people as you wish, but at least two. Compare the experiences you had of each.

THE EPISTEMOLOGY WITHIN which a person lives is communicated in every gesture, intonation, movement of eye and hand, every piece of clothing and stride of foot. It is possible, with practice, to learn to perceive all the elements of their epistemology, of their world, to know what it is like to live within it. To understand how other people experience them. To understand the emotional tenor of their lives.

It seems to me that those of us who work to preserve wild nature must work as well for a return to our senses and for a renewed respect for sensorial modes of knowing. For the senses are our most immediate access to the more-than-human natural world. The eyes, the ears, the nostrils catching faint whiffs of sea-salt on the breeze, the fingertips grazing the smooth bark of a madrone, this porous skin rippling with chills at the felt presence of another animal—our bodily senses bring us into relation with the breathing Earth at every moment.

—David Abram, "Trust Your Senses"

Exercise 3

Go to a place in nature that you like. (Be sure and take a journal with you.) Choose a place you have been to before. Find the area you like most and relax. Sit if you want to; get comfortable.

How does this place feel? Try to describe it in words. Be as specific as you can. Go on in your journal at length if you need to. Write down everything that comes to you no matter how silly it sounds. Even if you think it's crazy.

When you are done, allow your eye to rove, to be drawn to whatever one thing is most interesting to you. Look at it. Let your eye explore it, noticing everything about it. The colors, the shape, how it rests or grows in the ground. Its relation to the air around it, to the plants, water, soil, and rocks around it.

What feelings do you have? Write them down.

Is there any part of what you are looking at that you like more? Less? Why? Can you tell? Do all parts of what you are looking at generate the same emotion? Different emotions? Write everything down in your journal.

Do this with at least two other things that you see. Make sure that one of them is a plant. You can get up close if you want to, place your eye on a level plane, take an insect view. How is the plant shaped, how does it feel to your fingers, how does it smell? What emotions does it generate in you? Write everything down.

Only to him who stands where the barley stands and listens well, will it speak and tell, for his sake, what man is.

—Masanobu Fukuoka, THE NATURAL WAY

NOW, GO TO ANOTHER NATURAL PLACE, different from the first. Sit down and relax. Get comfortable. How does this place feel? Write down everything that you notice. Go on at length.

Does this second place feel different from the first place you sat? How are the feelings different? Which place feels better—the first or the second? Is there a name you can give the feeling you had at the first place? A name you can give the second? Names that will make clear the difference in feeling that you perceive? If you can't think of a word make something up.

This song of the waters is audible to every ear, but there is other music in these hills, by no means audible to all. To hear even a few notes of it you must first live here for a long time, and you must know the speech of hills and rivers. Then on a still night, when the campfire is low and the Pleiades have climbed over rimrocks, sit quietly and listen for a wolf to howl, and think hard of everything you have seen and tried to understand. Then you may hear it—a vast pulsing harmony—its score inscribed on a thousand hills, its notes the lives and deaths of plants and animals, its rhythms spanning the seconds and the centuries.

—Aldo Leopold, A Sand County Almanac

When you are finished, as you did last time, find something your eye is drawn to and write down everything that you feel and perceive. Do this as well with two other things, at least one of them a plant.

Each place on Earth has unique feelings associated with it, as does each thing that grows or resides there. The numbers of shadings of their emotional nuances run into the thousands. Each can fit into a specific space within the different human beings that need them. There is a richness in feeling, a companionability that comes from perceiving, the complex interweaving of emotional textures that reside in the life that surrounds us.

Going Deeper

Exercise 4

(Sometimes it is helpful to make a tape recording of this exercise and play it back. Instead of the words "them" or "they" that I use in the exercise, use "him" or "her" and "she" or "he" and "his" or "hers"—whichever gender you are. If you practice you will find the perfect speed, pitch, and intonation for you to listen to.)

Sit someplace comfortable. Someplace you won't be disturbed for a while. Someplace safe.

Close your eyes and take some deep breaths. Fill up your lungs as if they were balloons; fill them to bursting. Hold it a minute, then . . . slowly . . . release. As you let out the air in your lungs, let any tension you feel inside you release and go out with your breath. Do this again . . . several times.

Now. Imagine the floor or chair under you as two huge, cupped hands that are holding you. Let yourself relax into them. There is no need to hold yourself up; let yourself be supported.

Keep breathing and letting any tension in your body go.

SEE, STANDING IN FRONT OF YOU, the little child that you were. Notice everything about them. How are they dressed? How does their face look? Happy? Sad?

Are you happy to see them? Do they seem happy to see you? Will they look you in the eye? Do you feel comfortable seeing them?

Notice everything about your child.

Now. (Just inside yourself), ask your child if there is anything they wish to tell you. Listen carefully. Is there anything you wish to tell your child? Talk and listen as much as necessary until everything has been said. Is there anything your child needs from you? Is there anything you need from your child?

Now. (JUST INSIDE YOURSELF), ask your child if they will give you a hug. If the answer is yes, hold your arms out in front of you (actually do this), pick your child up and bring them to you and hold them tight. Let your arms go around yourself and hold tightly. Relax into it, feel what it is like to hold this part of yourself so closely. Is there anything else you need to say to your child? Anything your child needs to say to you? Let yourself be with this experience as long as you want to. Then, when you are finished, thank your child for hugging, for coming to see you, and, for now, say good bye.

WE ARE OFTEN TAUGHT in the Western world, especially in the United States, to divorce ourselves from this part of our self. It is a part of our self that feels very deeply and is very sensitive to the emotional nuances in the world. People often have difficulty in reclaiming this part of themselves. If you imagine a close friend whom you stood up three or four times in a row for a lunch date, you can imagine the level of feelings that might exist in a part of you closeted away for fifteen or twenty years. Sometimes it takes a great deal of work to reestablish communication.

[Luther Burbank] has the rarest of gifts, the receptive spirit of a child. When plants talk to him, he listens. Only a wise child can understand the language of plants and flowers.

—Helen Keller (in Tompkins and Bird, The Secret Life of Plants)

This part of human beings does not respond well to ultimatums or threats, but will sometimes respond to promises, especially if they are kept. (Usually you will have to do something in exchange. It is very important that you do it if you agree to.) It is worth the work it entails. Opening the door to this part of you opens the door to reconnection to the world and all the subtle meanings within it. I often suggest that people do this exercise daily for at least a year. This part of yourself will tell you everything that is going on internally, everything that you deeply need. It will also tell you much about the world around you. It is possible to become your own best friend. It is interesting that Luther Burbank, George Washington Carver, Helen Keller, and a great many indigenous plant peoples were all said to be like children.

Exercise 5

You can repeat the last exercise, if you wish, with any developmental age you have lived through, from infancy on. Each has its own intelligence, its own special connections with the world. Developmental stages do not stop at age twelve or sixteen; the child naturally grows to forty and to eighty. It is possible to remain filled with feeling and wonder and openness at any age. Each age has its own teachings. Each is a unique developmental stage of a human being's growth through life. Each has unique perceptions and capacities that aid in the experience of the human condition—at least they do when allowed to bloom, to grow unstunted and unrepressed. The early infant part of ourselves has the capacity to perceive everything simultaneously as it happens. Infants have no words (they perceive in gestalts), but that is all right; the child knows lots of words. And they are often willing to act as interpreter.

Exercise 6

Repeat Exercise 1. Go to the same places. Ask your child to be present with you, perhaps standing beside you and invisibly holding your hand. Let yourself relax and really begin to *see* and *feel* the store you are looking at once again. How does it feel to you today? Remember everything you know about it. Now. Ask your child what he or she feels. What part of the store feels best to them? What part do they like most? Ask them to tell you everything they are willing to tell you about the store. Spend as much time as they need you to spend. Are there any differences from when you went alone? What are they? Is there any pattern to the differences?

Go to at least one more place you went to the first time and repeat the exercise. What are your child's feelings and perceptions? Which place do they like better? Why? When you are ready to stop, make sure that before you do you thank your child for helping you.

Exercise 7

Now, repeat Exercise 2. Take your child with you again. How do they feel about this place? When you are seated and comfortable, begin

---

People would expect the most knowledgeable person on soil to be the soil scientist. But if, in spite of his extensive knowledge on soil as mineral matter in flasks and test tubes, his research does not allow him to know the joy of lying on the ground under the sun, he cannot be said to know anything about the soil. The soil that he knows is a discreet, isolated part of a whole. The only complete and whole soil is natural soil before it is broken down and analyzed, and it is the infant and child who best know, in their ingenuous way, what truly natural soil is.

—Masanobu Fukuoka, The Natural Way of Farming

looking at more people. Pick one that your child is most interested in. Have them tell you everything they perceive about that person. When you are done, have your child pick someone else. If they are willing to, someone they are uncomfortable with. Have them tell you why. What is it about that person that is uncomfortable? Have your child go into as much detail as possible.

Exercise 8

> *Go to the pine if you want to learn about the pine,*
> *or to the bamboo*
> *if you want to learn about the bamboo.*
>
> —Basho

Repeat Exercise 3. Go to the same place in nature you went before. Remember to take a journal. Find and sit in the same place. Let yourself relax. Imagine the Earth upon which you are sitting to be huge hands holding and supporting you. Take some deep breaths.

Now. Ask your child to come and sit with you. Have them tell you everything about this place. Go to the plant you sat with before. Touch it, smell it. Have your child touch and smell it. Have them tell you everything they know about it. Write it all down.

Now. Let them choose another plant—one they feel drawn to. Have them tell you everything about it. Write it all down. Repeat this at the second place you went before. When you are done, make sure you thank your child for helping you before you stop.

> *We must begin to develop the consciousness that everything has equal rights because existence itself is equal. In other words, we are all here: trees, people, snakes, alike. We must realize that even tiny insects in the South American jungle know how to make plastic, for instance; they have simply chosen not to cover the Earth with it.*
>
> —Alice Walker, LIVING BY THE WORD

SOMETIMES, LATER, IT HELPS to go and look up things about that plant in a book, perhaps a medicinal herb guide. The depth of information that the child can often gather from plants is amazing. People who have done these exercises with me over the years have described in detail information about plants they do not know and have never seen before. They have described medicinal uses, craft uses, clothing and building uses that are exceptionally sophisticated and are not apparent from the exterior appearance of the plant. I have even put plants in a closed box and have heard a person's child describe them in detail when the person, themself, could not do so earlier. It seems amazing, but it is not. It is just the way things are.

Exercise 9

Do these exercises often (you may even discover others of your own). The more you do them, the more connected you will be to your self, the more your capacity for perceiving and responding to subtle emotions will increase, and the more you will be connected to the world around you. After a while it becomes instinctual; this part of you integrates, is no longer separated out (either internally or as an exercise). It can take a long time.

As you develop this experience there will be a constant flow of information and complex feeling between you and the world in which you live. Smells may become vivid, colors enhanced. You may begin to take on a childlike demeanor, to spend hours sitting under trees talking with flowers. You may find yourself wearing bright colors and having a tendency to hum odd tunes. You many find yourself laughing for no reason or even gathering knowledge that you cannot explain with your rational mind. If you do this long enough and often enough you will start to have unusual adventures, you will begin living in a world where biophilia is commonplace. You will begin reinhabiting your interbeing with the world. You will start moving into biognosis. If you go on with it

[W]e are parts of a living world. . . . [but] most of us have lost that sense of unity of biosphere and humanity which would bind and reassure us all with an affirmation of beauty.

—Gregory Bateson, MIND AND NATURE

Integrity—the state of being whole and undivided.

—READER'S DIGEST GREAT ENCYCLOPEDIC DICTIONARY

even longer you might even realize that you are here to do something in particular, that you were born for a reason.

Why should we two ever want to part?
Just as the leaf of the water rhubarb lives floating on the water,

we live as the great one and little one.

As the owl opens his eyes all night to the moon,
we live as the great one and little one.

This love between us goes back to the first humans;
it cannot be annihilated.

Here is Kabir's idea: as the river gives itself into the ocean,
what is inside me moves inside you.

—Kabir, The Kabir Book (versions by Robert Bly)[6]

Living Biognosis: The Work of Carol McGrath, Sparrow, Rosemary Gladstar, John Seed

11

If it is the highest and the greatest that you seek,
the plant can direct you.
Strive to become through your will,
what without will, it is.

—Goethe

Oh, I understand! You were being a bridge.
Well, that's nice, bridges are important.
But you know, the only problem with being a bridge
is that you, yourself,
never get to cross over.

—Nan DeGrove

One of the reasons, I think, that the Gaia hypothesis has been so strongly seized upon by so many "lay" people is that it gives them permission to acknowledge something they have long known to be true: that Earth and nature are alive and that we are not alone, nor here for ourselves alone. Many scientists, professors, and scholars, even those who accept Gaia as accurate, have been disturbed by this popularization of the idea. It is as if only the formally educated can use the information wisely, understand it, or develop it further, as if it is somehow dangerous in the hands of the "uneducated." Lack of degrees or formal schooling routinely has been used as a reason to denigrate such popular sentiments, to assume that one can understand or do little about and for Earth unless one possesses them. This is a mistake. Biophilia always seeks to emerge. It will always break through sidewalks, those within *and* without us.

Biophilia and biognosis are natural capacities of all human beings. It is not necessary to formally study, to have advanced degrees, to be in established institutions in order to develop, hone, use, or refine biophilia or deepen it into biognosis. In fact, there is often an inverse relationship between these capacities and the possession of degrees. Formal schooling, as I eventually learned, has little to do with education. Sophisticated information about Earth and the interrelationship of its ecosystems and biota is available to anyone. *Anyone* can learn to understand the language of plants. All that is necessary is the capacity to feel, the willingness to observe in minute detail anything in nature that captures your attention, and the drive to reflect on what is felt and observed. You, yourself, can do this if you wish. If you so choose, you join a nation unbounded by time or geography whose citizens you may meet in any circumstance, a nation whose citizens are interwoven within the life web of Earth and whose work is fostering that life and its intelligence.

It is only in the mass movement of caring by many people, not the ponderous deliberations of the degreed, that care for the Earth will return to our species. Thinking will never restore caring. No matter how elegant the theory, the territory must still be entered and experienced. It is deeply ironic that one of the most powerful antibiotics Alexander Fleming ever discovered is in human tears.

The four people whose writing follows in this chapter speak of the work that their capacity for feeling and their deep connection to Earth has brought to them. Each has had tremendous impacts on the restoration and protection of plants, ecosystems, and folk and native peoples, and their ways of healing throughout the world.

Carol McGrath is a community herbalist and plant teacher who lives in British Columbia, where she has been working to restore her traditional Celtic herbal and Earth traditions. She carries on a long tradition of training community herbalists in the use of plant medicines, listening to the plants, and understanding the voice of the Earth. She is working to restore broken connections between peoples, people and plants, and people and the Earth.

Sparrow, though born in Berkeley in 1970, lived throughout his childhood in Ecuador. He returned to California to study for two years at Humboldt State College with Bill Devall and other deep ecologists, became active in the efforts to save the redwoods, and in a dream realized he should return to the Amazon, which he did when barely twenty years old. For the past decade, he has worked to protect and demarcate indigenous lands to protect them from logging; his main focus is the protection of indigenous medicinal plant knowledge and healing.

Rosemary Gladstar has been one of the primary forces for the restoration of American herbalism for nearly thirty years; she is thought by many to be the godmother of the American herbal renaissance. She formed the first new herbal school in the United States after the ascent of reductionistic medicine and for more than a decade she has supported the sharing of world herbal traditions by bringing herbalists from all countries together at the International Herbal Symposium outside Boston, Massachusetts. In 1990 she founded United Plant Savers to protect and restore populations of endangered North American medicinal plants.

John Seed is a founder of the Rainforest Information Centre in Australia and codeveloper (with Joanna Macy) of the Council of All

Beings workshops. He has been active in protecting ancient forests throughout the world and organizing the demarcation efforts in South America to protect indigenous lands (basically, cutting a 15-foot-wide circle around one and a half million acres of indigenous lands to protect them from development). He is one of the preeminent deep ecologists and Earth activists of our time.

These people did not develop their work in colleges, nor receive their education in school. Instead, they joined a long tradition of those who have reinhabited their interbeing with the life web of Earth and learned to listen to what the plants and Earth are telling them. Their work has emerged from this intimacy, this communication. Look deep, through their words and gestures and communications, and you will see the face of Gaia.

> Seeking for truth I considered within myself that if there were no teachers of medicine in this world, how would I set to learn the art? Not otherwise than in the great book of nature, written with the finger of God. I am accused and denounced for not having entered in at the right door of the art. But which is the right one? Galen, Avicenna, Mesue, Rhais, or honest nature? Through this last door I entered, and the light of nature, and no apothecary's lamp directed me on my way.
>
> —Paracelsus

SCOIL COISH CLAI: BY CAROL MCGRATH

I am a herbalist. Becoming so has led me not only Outward but Inward, Backward and Forward, until I am surrounded by herbs and life on every side. Whoever I was before has changed irrevocably. "Earth is our mother" is no longer a tired cliche I've heard too many times, but a living reality. It has made me aware of how wide the gap is between the attitudes I was taught as a descendant of colonial people, and the teachings I have since been given by natives and other Earth-oriented people, the plants that I love, and the land upon which I live.

In the early days of European–North American contact the colonial authorities quickly recognized that in order to gain influence in the "new" territories they would have to break the power of the medicine people on whom native people depended for healing and spiritual counselling. This, despite the fact that the colonists themselves needed and were often freely given medical information by these same medicine people. However, anyone as directly dependent on nature as were the colonial people *must* become observant of the natural environment—water, wind, trees, soil—in order to survive. We begin to notice when the deer come to drink, where and when particular beneficial plants grow, where certain birds lay their eggs . . . and slowly counter-acculturation occurs, we start to understand *why* the indigenous people of a certain environment have many of the customs they do. We begin to realize that our true roots begin with the land itself, that the environment of a particular place shapes cultural traditions and language, and that what we bring from one place may not be suitable to another.

The issues we face, living as the descendants of colonizers and colonized, have more to do with the attitudes of takeover and control inherent in the colonizing process itself, what I call the Colonizing Mind, than they have to do with any physical or racial differences we may have. We need to learn to see this Colonizing Mind wherever it may be operating, whether it's a government cutting down trees, a pharmaceutical company allowing water, soil, and air in a factory town to be polluted so it can keep its profits up, or a colleague telling us we need to make all our herbal tinctures "official." We need to constantly ask: "Who benefits?" so we know what the real agenda may be. The herbal movement can make a huge contribution to healing the wounds made to the world by those whose principles are based on self-interest and who have not realized that in a world where all of life is an interconnected web, Self eventually includes All.

Our roots as herbalists reach back through all ages, all places, and all peoples. Thus we have an opportunity to move into the centre of the web where we may learn from every part of nature: All people were once tribal people, living on the land, paying attention to the rhythms of nature. When we align ourselves with those rhythms and with the people who choose to live within those rhythms, we find the heart of life. This alignment is part of my story of coming to an awareness of the Colonizing Mind, and of the people who bit by bit drew me away from it and into the web of nature. I have been given songs, stories, food, herbal information, rituals and invisible gifts untold by people who were of a different cultural background than I am. The greatest gift was the key to my own culture, which came to be a key to the world of herbs and the world of all of nature. I would like to reciprocate, but the gift may have been too large. . . .

More than twenty years ago, shortly after I moved from an island on the East Coast to this island on the west coast of Canada, I was invited by a friend to go up island to a native longhouse, a "Big House," to help cook for an all-native conference.

"Call me Auntie Han!" said the smiling grandmother. Her warm eyes twinkled as she took my outstretched hand in both of hers and patted me on the shoulder. I immediately felt at home.

Although I was born thousands of miles from here I knew these ways. They were the ways of my people: warm welcomes, convivial companionship, island living, fishing people, and big families. Everyone seemed related to everyone else and kinship bonds were often referred to: "That's Maggie's daughter Sarah's boyfriend's nephew Sam." The kitchen was filled with warmth of fire and warmth of busy hands. The women and a few men moved easily around each other as we cut fish, peeled potatoes, mixed frybread, washed pots, and set tables. In between I went to sit in the main room to listen to speeches. Two big fires were burning in pits on the earthen floor, the smoke irritating to my eyes. There was drumming and singing, introductions and welcomings.

Eventually, the room hushed and a small, fragile-looking woman walked out and started to speak. Sitting on the hard bench I felt electricity go through me. She spoke Coast Salish, gesturing with her hands, yet somehow I understood everything she said; her words rang in my bones. She told the people about the importance of keeping the native language, pleading with the old ones to teach their children and grandchildren, encouraging the young ones to make the effort to learn. She spoke of how the spirit of a people is conveyed in the sounds of the mother tongue. Tears rolled down my cheeks and I put my face down, trying to discreetly cover my weeping eyes with my hand. I felt moved and embarrassed. It seemed as if I was listening to my own great-great-grandmother. It was only then that I truly realized how much I had in common with these people. Where they are now, fighting for their language, customs, children, spiritual practices, land, and lives, is where my people were four or five generations ago. For the first time I felt the damage that is done by the loss of one's native language. Language arises partly out of the need to express one's self. If we don't have the words with which to express who we are, we are silenced, cut off from our roots.

"SCOIL COISH CLAI." . . . I try to read the words written on the piece of paper my sister has thrust at me. The unfamiliar language tangles across my tongue, and my sister immediately corrects my awkward pronunciation, drawing the sounds out slowly for me . . . "Skull . . . cosh . . . clay." She is the only member of my family who has studied the old tongue, the language of our ancestors. "Scoil coish clai. . . It means Hedgerow School."

In the old days in Ireland, she says, our people were forbidden to speak their own language and were forced, by threat of imprisonment, to speak English. The old men and women had secret meeting places outdoors where they would hold classes, teaching the young ones native language and customs. Hedgerow schools. Moving constantly, hidden from the prying, spying eyes of the interlopers by the safe cover given by bush and tree, the ancient knowledge was passed on—knowledge of the livingness of Earth and our relation to it.

My Celtic ancestors believed that the life contained within us wasn't limited to our human experiences but extended to the experiences we had when we were other life-forms: animals, birds, fish, stones, or plants. These were all considered living beings capable of independent action and thought. The wise people were expected to be able to reach back not only into the memories of their current life, but also into their memories of all their other lives. Colonization changed that; I was cut off from my roots: I lived a childhood being told that creatures and plants were alive but only humans had real souls. It was when I came to the West Coast and met people of the Coast Salish tribe that I heard stories of metamorphosis that were like the stories my people shared centuries ago, stories that indicated that humans are not the peak of evolution but part of a shifting interchangeable web of life.

I like the idea of an outdoor school with no walls or ceilings. Within the Celtic tradition all of the outdoors was a place of teaching. But to be cut off from our roots means to be cut off from the outdoors, from our culture, from our very nature.

In true scientific fashion, let's look to the lab rat for understanding. Instead of the variable warmth and textured surface of a wild home nest, it has a constant temperature and a stiff, unyielding cage floor. Instead of a variety of smells changing according to the time of day, weather, and season, it has the chemical scents of the laboratory. Instead of the kinesthetic experience of a social animal that lives in intimate physical contact with other members of its family, it is often placed in isolation. "Lab chow," hard, dry, and lacking in multiple chemistries, replaces the complex foods of various textures, moisture levels, and content that it would eat in the wild. And what about the absence of wind, rain, and full-spectrum light? Suddenly the "truth" of objective scientific data becomes curiously distorted.

Thinking of the lab rat I am reminded how far we have come from our ancestors' practices of living directly with the natural world. As I move deeper into the world of herbs I have begun to sort out differences that I wasn't immediately aware of. First: I find that when I go out to gather the plants, my walking on Earth and stones exercises different body muscles than walking on sidewalks. If I go barefoot, my awareness increases: There is always the possibility of sharp or mucky things underfoot. My muscles, freed from the repetition of an even surface, develop differently and my body shape begins to change. Then, there are the smells. When we smell things we are ingesting small amounts of whatever might be in the air. In the woods on a warm day I breathe in a mixture of the essential oils of trees such as Douglas fir, red alder, grand fir, and black cottonwood; aromatic shrubs; and wild herbs. These plants all possess medicinal properties, and when we camp or live among them we take in small amounts of their antibacterial, antiviral, antifungal medicine all day and night. In the city, I smell car exhaust, tarmac, paint, plastic, felt pens, foam cushions, floor cleaner, people's deodorant. This is not food, it doesn't nourish, and yet my body takes it in because that is what bodies do. Ingestion of these substances is a new experience for the body to deal with and the rise of chronic disease seems to indicate it may not be a particularly successful evolutionary experiment. These are new substances introduced into a traditional food chain. No one *wants* to eat aluminum, formaldehyde, butylated hydroxide, polyester fibres, or plastic molecules mixed in with food, but the new way of life creeps up on us and before we know it we're colonized, trapped in the plans of others. Every factor of a healthy or unhealthy existence that the rat must deal with, we also must deal with. Our way of life is wrapped around us like swaddling clothes, so tight it hinders us from getting outside where we can get a good look at it. Carl Jung reportedly said, "How do you find a lion that has swallowed you?"

A friend of mine went to a women writer's conference and heard a native woman speak. She said, "If you take the Christian bible and put it out in the wind and the rain, soon the paper on which the words are printed will disintegrate and the words will be gone. Our bible *is* the wind and the rain." She suggested that people interested in the North American native tradition go back to the roots of their own native tribal traditions, back as far as is possible, before books, before the Christian tradition, to see what might be there. Cut off from my roots by time and language I followed her advice, and found my ancient cultural traditions a treasure chest of raggle-taggle talismans: a story here, a name there, a piece of a song, a yearly custom, an item handed down.

We all have these cultural remnants. They are a mother lode that can be gently mined to direct us to the riches of our ancestral lives. Sometimes information will come in less direct, less intellectual ways . . . through intuitions, urges, or as messages in dreams.

The idea of meeting someone in a dream before you actually meet them is not infrequent in indigenous medicine traditions all over the world. I discovered my people had many traditions similar to native North Americans, customs of spiritual quest and divination, rituals of purification, veneration of ancestral lore, and these included ways to approach the dream world. In many cases this information was handed on in an indirect fashion. I learned as a very small child that you could be awake and aware in a dream, that you could change your dreams at will, that you could "program" your dreams to be about certain subjects. My father told us about a dream he'd had as a boy. He dreamt he was walking up a hill with a basket of apples over his arm, when the friend who was accompanying him said, "Jimmy, let's sit down on these steps here and eat some of these apples!" My father replied, "Okay, but we'll have to eat 'em fast, because this is a dream and I might wake up!"

This story always made us laugh because of the boys' enthusiastic eagerness for the apples, and also because of the magic involved: that a person could be sound asleep, fully aware of the fact, and making decisions based on that awareness. When I began to learn about plants from native people, this story made it easy for me to follow some of the dream practices that related to gathering information about plants and about the people who needed them. First, you have to believe that the dream world is meaningful. To native people the dream world is *very* meaningful. Information that comes in a dream is considered as important or even more important than what might come during regular waking time, because in a dream you are interacting directly with the spirit world. (Some may find it easier to think of it as interacting directly with the world of the unconscious or subconscious). The idea is to access material that in our normal waking state may be closed off to us. Often a native medicine person, before seeing a sick person, might go to sleep to seek guidance from the spirit world about what treatment to follow.

In Celtic tradition, similar practices existed, sometimes associated with particular mythological figures: Gwynn, god of the underworld, king of Faery, or Helen of the Roads, goddess of twilight and dreaming. Gwynn is equated with the Romano-British figure Nodens, god of the abyss and a guardian of the woods. In a temple dedicated to him in England, excavation showed facilities for incubation, places where people slept in the hope that Gwynn would come to them in a dream and give them a message. This figure's connection to the deep inner realms on the one hand, and to the woodlands on the other, is not surprising. A deep bond with the natural world, and taking time to observe the signs of nature, are prerequisites for the development of deep understandings, for precognitive or prophetic visions.

I HAVE A DREAM ONE NIGHT . . . I am gathered with a group of friends at a Sunday ceremony led by a herbalist. He speaks of the plants, of the spirit that lives in all things, of the need for restoration of the Earth. The ceremony doesn't take place in a church or even in a beautiful pastoral setting. It's in a vacant lot in the inner city. We are surrounded by the potsherds of modern civilization: old tin cans, broken glass, worn tires. The faces of the people around me are shining as they listen to his words. We hold hands and sing. Then we pick up all the garbage, sorting it into piles. Some will be recycled, some will be used for other purposes, and some will be moved from this place to a central depot where it can be processed or buried.

A week after I have this dream I invite a group of women to come with me to a nearby hill, triangular shaped, which has been cut off, isolated, by highways all around. At the bottom of one side of the hill is a bus shelter and litter blows up the hill from the bus stop. It's springtime and hundreds of flowers are coming up, squeezing out from under chunks of rusty iron and old bits of plastic. We spend a couple of hours picking up garbage and just as we finish an elderly man climbs the hill and asks us what we're doing. He tells us the history of the hill and surrounding areas. When he was a boy he delivered newspapers to the thirteen houses that were on the hill back then. He shows us the area where a man kept exotic animals to overwinter. He'd hear the lions and tigers roaring, the elephants trumpeting, as he did his paper deliveries. He gives us the names of some of the plants there that we aren't familiar with, and eventually he goes away, seemingly puzzled as to why we are bothering to clean up the hill, since we don't live next to it as he does—but he is delighted all the same. We sit in a circle together after he leaves, we sing, we give thanks for his gift of old-time stories about this place. This place where he belongs.

I was raised in the "New" World, speaking the language of the people who had colonized the ancestral territories, the descendant of colonized people who had become, in their turn, colonizers. My family has lived in this "new" world for more than 200 years. Where I grew up, people would ask not, "Where are you from?" but, "Where do you belong ?" It's a fine and important distinction, as if your place of origin can own you as much as you own it. As if, despite whatever cultural changes occur, there is always the land itself and the things that rise up out of the land: rocks, rivers, wind, rain, and all varieties of plants. As if it's natural order, not cultural order, that ultimately gives us true security. To some extent that's true, no matter what information about plants has been lost: We can always go back to the plants themselves to get the truth. They will always know what they are and what they can do.

It is sad that the first thing we learn about plants as children is often "No!" We learn that nature is dangerous, that wild plants can poison us and that we shouldn't put them in our mouths. We learn that a few dozen plants such as carrots, potatoes, and culinary herbs are exceptions to the rule, and for most people that's where it stops. But the Earth community itself can speak to us and teach us more.

Once, a medicine man was coming to visit my home. The day before I spent the day cleaning, helped by the young daughter of a friend. Late that evening we decided it was such a beautiful starry night that we'd get our blankets and sleep outside, but first we'd have a calming cup of herbal tea. I had a small patch of chamomile that I had grown that year for the first time and had never picked. I went out to pick some and brought it into the house. The petals were curled up tightly, as they do at night, and we made a joke about it, "Ahh, how sweet, they're sleeping," and proceeded to make the tea. The next day I was showing the medicine man around the garden. We talked about the different plants, and unexpectedly, as we strolled by the chamomile patch, without even looking at it, he turned around and shook his finger in my face, saying in a suddenly raised voice, ". . . *And you should never pick them while they're sleeping!*" My jaw dropped. Startled, I stammered that I hadn't known that. He replied in a perfectly normal tone of voice, "Oh yes, plants need their sleep, just like people."

This idea of plants having needs "just like people" is one I hear over and over again from native people. I have been told: When you go out to pick the plants, you must start talking to them before you even leave your house. You must ask them for their help. Tell them why you need them. Ask them to reveal themselves to you. If you are picking plants for a particular person, name that person, and tell the plant what the person's problem is. If you pick for a certain person don't ever use that medicine for someone else. If you don't know exactly who will be using the plants, just say that it will be helping somebody. Don't ever pick the first plant you see, but leave an offering there—it might be a prayer, a song, a simple thank you, or something material like a piece of your hair, or some tobacco or sagebrush. It's the intention that's important, the acknowledgment that the plant is giving itself for your benefit. Respect for that gift must be shown.

Another time I was told, "It's not simply that you need to learn about the herb, but the herb needs to learn about *you*. You must share with it who you are, what you're up to. That way it will come to know you and will be better able to help you in your work."

One year I went to Central America and I heard the same thing from a herbalist I met in the marketplace in Belize. She said, "I was taught from the age of seven until about thirteen by a little black man, what you'd call a pygmy, from Africa. He was more spirit than man, that man. He could hardly speak my language and I spoke none of his, but I always knew what he was telling me. He would take me into the jungle and make me dig those roots, and climb up the trees to get those leaves, and every single herb I had to talk to and I had to eat it myself so I would know what it was about. He would come knock, knock, knocking at my mother's door and when she answered he would say, "Give me the child!" and she'd hand me over to him and off we'd go into the jungle."

When I met this woman, she said to me, "I had a dream last night that someone special was going to come to talk to me about the herbs; you are the person who was in my dream!" She gave me a herb that she said was good for cancer, another that would give me energy, a third for when I needed to rest and relax, and a fourth that just "tasted good and will make you feel good."

I've thought many times about this woman and the other herbalists of many cultures who have shared their knowledge so freely with me, and I have questions about the attempts being made to "regulate" herbs, herbal practices, and herbal schools. I wonder about the scientific herbalists with their Thin Layer Chromatography and "standardization" and remember Deepak Chopra in *Quantum Healing* saying, "The first thing that is killed in a laboratory is the delicate web of intelligence that binds the body together." How can I learn to distinguish between progress on the one hand, and the Colonizing Mind creeping in to tell me where, when, and how to grow herbs on the other? If we cut off our roots, which grew out of the rich tradition of folk healing (which itself grew in the compost of the experiences of millions of people), will we still be able to stand?

MEDICINE > GARDEN > WHEEL: BY SPARROW

Sitting at the bow of our small canoe carved from the trunk of a wild cinnamon tree, we quietly drift downriver. There is something so mystical about being in the wildlands, within a deep forest extending uninterrupted for hundreds of miles. The loud ruckus of birds and insects, the occasional monkeys thrashing through the trees, the plopping of turtles off logs into the river, all add to this feeling I have of "being where we began."

Still, silently rowing downriver, I feel an isolation. I suppose the reason why I feel such a thing among all this life is because I am part of a people that have drifted from primal Earth.

Oh great soup of life, I swim through you for a brief second in your evolution. Yet you inspire me beyond passion and reason.

A peculiar flower hanging over the river breaks my melancholy. Hollering out my usual, "Stop the canoe," we frantically paddle against the swift currents. Juanchu swims out and grabs a few flowering specimens. "Kinoing? (What is it?)" I ask my companions. "Ome," they replied, meaning vine. The usual reply when it is too far away to identify or they don't know.

Juanchu swims back with specimens. A silent moment, everyone thinking "what is it?" Then the fragrant garlic plume hits us and almost simultaneously everyone cries out its name in their own language. "It's the garlic vine!" "Wiyayen!" "Waaponi!" What joy and excitement fill us all! And the excitement of our Huaorani friends feeds our excitement and ours feeds theirs. Our smiles grow. These forest vines are rarely seen in bloom. A fortunate sight indeed. "Jah! Give thanks and praises." We all holler and cheer. Here in front of us the legendary garlic vine, and in full bloom. What joy I feel to see my Huaorani companions, age-old friends with this forest lanai, so ecstatic. The connection is there, I think to myself, between plants and people, and their strong relationship that goes way back.

The garlic vine's lilac blooms drift scent now, low on the river like the morning fog. Large bright flowers fill the crowns of nearby trees, hundreds of them, bunched together, hanging down over the Rio Quehueiriono. Our canoe slowly drifts by and as we enter the fragrant plume we breath deeply, inhaling the fragrance of this sacred vine.

Oh medicine vine, there at the center of the sacred hoop, growing abundantly on the banks of the Quehueiriono, headwaters of the Amazon, we salute you.

Wiyayen, our Huaorani companions call it. It is *Sacha ajus* in Quichua, *Ajo de monte* in Spanish, *Kaip* in Shuar, garlic vine in English, and in Latin *Mansoa standleyi* (Steyerm.) A. Gentry, Ann. Mo.Bot.Gard. 66 (1980) 783. of the Bignoneaceae family.

This large, primary forest lanai is an old friend of many Amazonian people; its medicinal properties are part of many tribes' age-old knowledge. When crushed, the large, smooth, dark-green leaves smell strongly like garlic. And its gray-brown woody vine has swollen nodes as well as the strong smell, making it easy to identify when not in flower.

We've been drinking it often on our long journey pole-canoeing down this remote Amazonian river in eastern Ecuador, and consider it a companion, a friend giving a helping hand, along our exciting and adventurous journey.

Many Quichua-speaking people, Runa people, consider garlic vine to be effective toward preventing malaria, promoting good health, and staying strong. I have been told of its effectiveness in healing hepatitis, yellow fever, typhoid, common colds, coughs, fevers, sore muscles, and sore throats. The Huaorani use it also for muscle pains, stiff joints, and when the "inside of the bone hurts." The Shuar say it is effective to protect oneself from the spells of black magic brujos, and when planted around the house it is believed to keep away poisonous snakes.

A friend of mine once came down with typhoid and I witnessed his healing through the use of this vine. Its bark, cleaned and rasped, then mashed in hot water and drunk daily, brought him back to healthy living within three weeks. Since then I have grown to strongly respect this vine and regard it as kindred.

Once, on a long jungle journey, an old Huaorani man of wisdom noticed my extreme enthusiasm and happiness toward this plant. I had found a large vine, as thick as my leg, tangling through the towering canopy. I dropped on my knees praying, hugging this sacred plant, feeling the calm wisdom emanating from my silent friend. I sang, thanking the Spirit. Praying for it to maintain my strength, to protect me from the treacherous malaria looming in this country, I thanked it for being my friend and helper, and even climbed half way up it! We both sat there laughing, happy about our encounter with the vine. Then I carefully cleaned and rasped a few handfuls of its strong-smelling bark, wrapped it in a Geonoma palm leaf, and walked away content.

And the more we sang, the more we lost all doubt in the medicine.

In camp I prepared a thick, strong juice, offering it to my companions. Some drank and others refused because of its horrid smell and flavor. That evening, the old curandero told me while under his Jaguar trance that I would not get sick, nor get malaria, and at the end of the trip what he said became true. None of us had taken any modern malaria medications and we had all been living under the same conditions, yet my friends who had not taken a drink came down with it.

These events leave me with a strong love for this plant. Knowing a plant, its names and uses, is, when you find it again, like reconnecting with old friends. Once you know a plant, it is a friend who is faithful but not always there. The only time I've gotten malaria was when the garlic vine was nowhere to be found!

Our journey downriver continues and around nearly every bend of the river we see the garlic vine in full bloom. Every time we see it we all stand up, wave our hands in excitement, point and laugh. Its beautiful, large, shiny, lavender blooms, all clustered together, excrete the most exotic of sweet and sour fragrances and capture our souls.

Oh great forest, these smells, this breathing, these sniffs, these snorts. We rejoice in your fertility.

The connection is there. For some it goes way back, and for others it is so new. People and plants, a mutual energy flow as old as the forest and humanity itself.

I am a plant fanatic.

I get overjoyed when encountering familiar plants. So many times I have come across a plant that I have never seen, but its energy is so strong that I know it must be medicinal or special in some way. Medicinal herbs have a strong vibration and when in tune to this, you just know it's one of those plants that can help you as soon as you encounter it. Many times this has happened to me. Upon asking my indigenous friends what it is, they say, "Ah, this is a special plant. *Buen remedio* (good medicine)."

It's an amazing feeling to be in tune with plants this way, to feel such a part of the living Earth. At such times I feel the strong connections that exist not only between people, but between beings. Both alive, sharing the beauty of life together. In the forest, wandering through the towering trees or crawling under a thick tangle of vines, there's no need to feel alone. The forest is full of friends. I think how wonderful it would be to know the names of all these plants, every tree and shrub, moss and vine. To understand their dispersal, their pollination, which animals are involved in their lives and which plants enjoy living near each other. I suppose this quest for connections is yet another part of humankind's search to be a part of something greater, a part of my search to be in service to something greater than myself.

Living in western Amazonia, among the world's most biologically diverse regions, tribesfolk have an intimate understanding and familiarity with their forest home. Through observation, patience, acutely refined senses, and curiosity, these people have become at one with the forest. Some tribes have names for almost all of the flora and fauna in their territory. They have learned about pollination and dispersal and the interconnectedness between plants, animals, birds, and insects. These people have learned to read the forest book; it is the greatest encyclopedia, the most majestic university, and they all become its honored alumni. Cultural knowledge has been gained over generations as each tribe member learned a bit more. This is shared and the collective group knowledge grows.

Perhaps the most impressive and in-depth understanding of the values of plants can be found in the consciousness of Amazonia's western tribes. Living in a forest of immense biological diversity, where scientists estimate there are more than 50,000 species of plants, the tribal members of this extensive bioregion have become some of the world's master "deep ecologists" and "deep herbalists." (This is for you, lost amerika.)

These people are deep ecologists in that they intimately understand how humans are woven into the web of life. Hunting, gathering, and gardening techniques are involved, complex, and often ritualized. Taboos are strong and only certain animals can be hunted. What many consider savagery, brutal spearing raids or infanticide, may simply be their maintenance of the fragile balance.

They are deep herbalists because their understanding of the medicine in the forest is profound. They know how plants affect humans, and how to use plants to heal. Some tribal folk are such masters that they can establish contact with the spirit realm through the use of certain plants. They use plants for divination of illness, to predict the future accurately, to warn the tribe against enemy raids, and for learning where the largest game herds might be. Many medicine folk have been able to use certain plants to learn the uses and preparations of other plants, or to understand the intimate behavior of the most secretive animals.

They know, and I have learned, that an energy flow exists between people and plants. It can be looked upon like a yarn weaving us into the web of life, a direct connection that exists between people who know plants and their characters, and vice versa. Plants have people friends as people have plant friends. The flow is equally as strong from plants to people, yet much more subtle. We humans live in a realm a thousand times faster than plants and rarely do we slow down enough to hear them speak.

> *Slow down to you, slow down to you . . . slow . . . down . . . to . . . you . . . s l o w d o w n t o y o u . . . Plant Nation, we slow down to yoooouuuuuu . . .*

Passing down accumulated tribal knowledge has always been a central focus for indigenous people. It means the survival of their people and the continued maintenance of their territory. Through chants, dance, storytelling, hunting, gathering, gardening, usage of hallucinogenic plants in ceremonies, and constant contact between the elders and youth, tribal knowledge is passed on.

Until recently, indigenous people of western Amazonia remained largely uncontacted by outsiders. It is only in the past few hundred years or less that these tribes have been contacted, some only in the past thirty years, and some still remain uncontacted. Most tribes have not survived these contacts well, they have undergone shattering crashes and deculturalization. Missionaries and their schools have been influential in "knowledge collapse," pulling away the youth and turning them into new things.

Though Western medicines and medical services have saved many lives, they have also been used to demoralize the people and take away their pride and independence. Missionaries have often said that the garlic vine is the "devil."

With the recent industrialization of western Amazonia, the situation has grown more severe. Road building has brought waves of colonists into these "free lands." The colonists and their "work the land or you're lazy" attitude has been a powerful brainwash that has successfully crept into the lives of some indigenous groups, and has driven them to destroy their forest for cattle or cash crops lest they face losing their land. Although this is now changing with the legalization and demarcation of indigenous lands, to the colonists the rain forest is "empty" and free and government policies have enforced the principle that if you "work" the land you own the land.

In a certain sense the ball of destruction is rolling downhill, gaining momentum, and to some extent, out of national government's hands and into the jaws of the greedy corporate mega-complex . . . *hiss, growl, snarl, steam vents and steel.*

Whether slow or fast, the modern world and the rain forest are headed toward a shattering crash. Amazonia's forest masters are undergoing "knowledge extinction," even faster than the disappearance of the forest itself. Much knowledge has already been lost, but much still remains in the memory of the elders who have lived the old ways.

Times have changed. The youth go to schools and spend their time away from home.

Yet, in no matter what age or time, there always seems to be a special family of persons who are determined to maintain the dance. Here in Ecuador I have met some of these inspiring souls. Inspiration is like a wheel. One person, if inspired and determined to learn all they can and let it radiate, will inspire another. The wheel of inspiration, like that of destruction, has begun its downhill roll, and is gaining momentum.

Inspiration can start a chain reaction when the elements of the moment are just right. Inspiration is a force that often comes from some unconscious source, flowing into the conscious mind of a human and creating new ideas and ways of being. Inspiration chooses a medium through which to flow—it could be through art, a landscape, or a garden.

Medicine > Garden > Wheel

In Ecuador, around upper Napo province, especially around the capital city of Tena or Archidona, the Quichua people have had many centuries of contact with missionaries. This experience has strongly changed their traditional paradigm: from independent in the wilderness to being dependent on the modern world and its things. Cattle and cash crops have replaced their forest, and modern medicine has replaced many of their traditional medicinal plants. Many people have died from this loss of the plant connection. Western medicines are just too expensive or the doctors too far away.

Recently, there is a growing movement by the people themselves to recapture native knowledge before it disappears. The elements of the moment are now right; this is a process to support.

Models mean more than an infinite amount of words. At Jatun Sacha, a biological reserve on the upper Rio Napo in Amazonian Ecuador, we, at el Centro de Investigacion de Bosques Tropicales (CIBT) in collaboration with Fundacion Jatun Sacha, have created a model medicinal garden. This garden now harbors more than 150 species of the local medicinal flora. Seven years old now, it has been influential in inspiring three other similar gardens in the Archidona area. The concept is "Medicine—Garden—Wheel." The plants are the medicines, the garden is the tool. Together they are a wheel of inspiration we have started rolling downhill.

We hope to inspire many more medicinal plant gardens to support this growing movement to maintain sacred plant knowledge. The garden at Jatun Sacha also serves as a pioneering center for cultivation research. Many of the wild forest medicinals have never before been cultivated, and we hope to eventually distribute this vital information. For there is something important that happens with plants, something important that modern medicines can never do.

A phenomenon evolves when someone is healed by a plant. A connection begins and usually develops into deep respect and caring for that plant. When wandering through the forest, if one comes across one of these plant allies, one instantly feels the connection. A bond evolves between plant and person, a love grows, and it is shown by the enthusiasm expressed when seeing it in flower. This is how plants can weave us back into the web of life. Then once again we know our place. Once again are home. Not only are our bodies healed but our spirits as well.

Oh greatness i humble myself to you. we are all equal, we are all the same. we are all great. we are all human beings.

GREEN, I LOVE YOU GREEN: BY ROSEMARY GLADSTAR

Green, I love you Green
Green wind, Green branches

—Gabriel Garcia Lorca

It is a stark realization when one enters the desert, the wilderness of the soul. Inevitably, we all are called there. It happened to me when I was very young; I was eight.

We had moved from Wildwood Dairy, the place of my birth, to a small farm in the foothills of coastal California. Our new home was surrounded by remnants of ancient redwood groves that in bygone years covered the entire valley floor. I found temples growing there, undisturbed soils, small streams, native land. For the first time, I recognized sacredness in undisturbed landscape and it became part of me. After nearly five decades, I can breathe out, and remember the first time I walked in an 800-year-old forest.

I like to wonder occasionally, futile as this exercise is, what my life would have been like if it hadn't happened. If I hadn't made friends with Joanne, the farmer's daughter. Hadn't wandered across the field to her house on many occasions, and on this one, wandered into the barn where the farmer was. The musty dark barn. The cold gray milkhouse. The end of one life, the beginning of another.

I can only remember fragments of what happened. I have ceased wondering much at all. But what I do know is that my life was radically changed: I could feel a searing pain, located somewhere beneath the breastplates protecting my heart. Oddly, even at this young age, it awakened in me the compassion that sometimes comes with knowing such pain. I felt more, and that, more deeply. I was more sensitized and more open. I was also more closed. And yet, in that moment, I could feel the plants that had surrounded me since childhood reach in to that place that was sealing off and begin to heal the wound.

My favorite plant at that time—it is still one of my best friends—was a small, wild mint found growing along stream banks, in open meadows, and on unmowed lawns. Self-Heal. Aptly named. It is a remarkable plant, healing on many levels, from colds to infections to souls.

After, I would lay on the grass, surrounded by these purple blossoms. I would fall asleep, fully immersed in their essence, feeling them lightly caressing my body. They had a soft voice that seemed to sing the gentlest of songs, offering a healing balm to a bruised being. They were truly a plant ally for me long before I knew the word for such a friend.

THESE ARE THE TWO DREAMS that came to me at this time of despair. The plants were entering my unconscious, healing me in places no other medicine could reach.

I am running, lightly, joyfully, across a path lined with small purple violets, all nodding to me, fully surrounding and embracing me. A radiant light is flowing through an enveloping mist drenching the plants with joy. My wound fills with this radiance and my heart begins to soar again. Violets, I learned later, are for strengthening the heart and lifting the spirits.

The second dream was a waking dream, one that recurred many times in the next few months. It became a vision dream, a directing force in my life. I was visiting my favorite oak tree in the field behind our farmhouse. It was already several decades old at this time when I was barely nine. I would go out to it, this huge Valley Oak, one of a long line of people, most of whom looked very sad or crippled in some way, all of whom had journeyed to this elder to be touched. I would stretch up my arms to the old man who sat up in the limbs of the tree and be lifted high. He was very old, ancient, with a long thin white beard and, surprisingly, he was completely naked. Not a stitch of anything on. He seemed unaware of this and no one else, including myself, seemed to notice anything odd in it. I never questioned how any of these quiet, bent-looking people got there or what they were doing in this field on my parents' farm. Nor did I question what that old gnarly figure was doing up in the tree. It all seemed perfectly normal and perfectly sensible that we should be seeking him for healing. Oak, I learned later, is the medicine of enduring strength and offers hope to those in dire need.

TODAY, THOSE MAGNIFICENT VALLEY OAKS are in grave danger and reach out to us for help. Suffering from an "unknown" disease, these magnificent trees, when smitten, often die within months. My life and my life's vision are intricately connected to that oak community, so I have returned westward to the valley several times these past couple of years to bear witness. I have little in the way of science to offer, no pest controls or magic cures, but I have a profound relationship with the spirit of those oaks, and a great desire to help them. When I walk out to talk to them, I pray for their spirits, for their lives, for their connectedness with all things, and for the humans who are bringing this sickness.

I'm told very clearly by these elder trees that their roots are saturated with the sludge and discharge of the sewage plants that were positioned nearby to purify the city's waste. A green belt, basically a sewage discharge area, was created around the nearby city as a way to process these wastes. The discharge is purified, declared "almost" drinkable, and sprayed over this living green. It looks very lush, an emerald green carpet of grass and wild mustards watered by the city's waste. But the oaks are a dry species, thriving for centuries on the hot rainless summers, rejoicing in the winter rains. And no matter how many chemicals are used to purify the city's sewage, and how many chemical processes are then employed to eliminate the chemicals, it's not rainwater. The oaks call out to any who will listen; they speak endlessly of what it is they need.

Plants *do* talk to us and teach us. Their language is accessible, or I should say languages, as they speak in many tongues. It is not a lost language, nor is it accessible to only a few. Children hear it best, but are often discouraged. And like any language not spoken often, after awhile, it begins to fade. By middle school most children have forgotten the exquisite language of the plants. My life has been blessed; I did not forget the language of plants. My wounds and my upbringing made a place for their voices in my life.

My relatives said I was planted, not born, on this Earth. And it is true, the plants called to me. In my earliest memories I hear them quietly revealing their stories. I wandered in the meadows and woodlands surrounding our first home from the time I could walk. It was a good place to be planted. Hidden in a sea of grass, wildflowers woven within. I remember tasting my first wild mustards, cress, and dandelion greens. I knew I was being entered into. My family left me unattended enough so that wildness could root in the core of my being. While I only half-listened to the talk of people, I immersed myself fully in the language of the green. I became part of the underground network of the plants. I thought I was born in paradise and that all the world was a beautiful reflection of the verdency I knew in this place of my childhood.

My first "apprenticeship" in the deeper knowledge of plants, my real elementary school, was with my Armenian grandmother, Mary Egitkanoff, after whom I am partly named. My other grandmother, Rose, was a gentle woman, a wife and mother whom I never knew well. Mary, my mother's mother, was quite possibly the most remarkable woman I have ever known. A survivor of the Armenian holocaust, she was steeped in the lore of herbs, and had a faith in God that was unshakable. She taught me, yes, but it was an unspoken apprenticeship. She did not "formally" train me, as she herself was not "formally" trained. She lived in a time when herbs were the medicine of choice. The remedies that she knew and used had been used by her mother, grandmother, and every other woman in her family far, far back in the family tree. She had tucked these remedies securely in the parts of her heart that survived the forced march through the killing fields and brought them with her, planting them in the hopeful soil of her new life. Planting them inside me. Thus I learned the commonest weeds, their names, their uses, and most importantly, their stories. Grandmother listened, too. She knew their language.

"*Oochias* (little rascal)," she would take me by the hand and say. "This is dandelion and this here is purslane. I ate this in the wilderness, on our march to death, and I lived. *See*, Oochias." And I saw and learned and remembered.

My second apprenticeship, my degree program, was with wilderness and wild plants.

I have just turned twenty and I have an infant son. With a friend, we head north to the mountains, the big ranges, to live as close to the land as possible. I know very little about living off the land, except for a growing knowledge of plants. I head right into one of the lushest zones on the North American continent, the Olympic Rain Forest. Surrounded by an intact primary forest that supports hundreds of bizarre-looking fungi, untold numbers of plants, elk, bear, deer, and other wild creatures, I begin my degree program. I spend two years living in what I now call "a state of grace," but which my parents called "out of her mind." I found little cabins in the woods, lean-tos with moss-covered beds. I stretched the levels of my comfort, walked barefoot through three seasons, and learned to receive nourishment on every level from the surrounding wilderness. I would lay on the ground, face close to the tip of a fir, and let the drops of water slowly filtering through the tree nourish me. I ate nothing but what I gathered: fungi, berries, lichen and fresh greens. I became what I ate. I was happy. My son was healthy. Our spirits were full.

I returned to California only long enough to earn money enough to follow the rest of the dream, to finish the degree. As in the storybooks of old, it was an early spring morning when we rode off "happily ever after." We were an odd little party, hardly what one would expect for an endurance trip: my son, Jason, who had just turned three; Evie, a close friend, who had volunteered to come with us, partly out of curiosity, I'm sure; myself; and two well-mannered but ill-prepared saddle horses. We headed due North, riding the coastal trail for a couple of hundred miles, then heading inland.

Our equipment was so simple. I laugh now as I prepare to head out on a family camping trip. Therma-rests, and hi-tech sleeping bags, special lightweight cooking utensils, a small propane stove, tarps, and tent all piled into the truck for a weekend a few miles from home. We were gone for four and a half months riding two amazing steeds with little gear other than a red plastic bucket, a couple of pots for cooking up the wild greens we collected along the way, a small Swiss army knife, three old J.C. Penney sleeping bags, very little clothing, and an abundance of youthful stamina. We rode every single day, resting from our northward trek only when the horses needed reshoeing or an extra feeding. Mile after mile, one foot in front of the other, we watched the trails, the hills, the coastline disappear as we traveled up the magnificent Lost Coast, up through the foothills, overland, to the Trinity Alps of Northern California.

We took no food with us other than a small bag of snacks that we would replenish whenever the rare opportunity presented itself. Our primary food source was the wild greens, berries, bark, and roots we collected along the way. Our very life depended on the generosity, the teachings, of the plants. They were under us, in us, around us, offering us their abundance in every turn of the trail. I became acutely aware that the air around me, the air we were breathing, was a divine gift from them. The food that became my body was of them. The beauty that surrounded me was their beauty; the essence of a hundred hues of green arranged upon the Earth's mantle became my essence. Even their sacred medicine was offered so freely, a gift to be used. When one morning we woke to find that Jason had been bitten by a large blood-sucking insect, his eye swollen shut, I was directed clearly toward a plant, told to pluck a few leaves, pulverize them, and place it as a poultice over his eye. It was no surprise to me that the swelling went down within a few minutes, and we prepared to ride on. Today, after reading so many books, listening to so many human voices, and warned so often of the side effects of plants, I have to stop to listen to hear plant voices speak again that clearly.

We never made it to Canada as we thought we would. Our day-by-day plodding, four and a half months of riding, got us only as far as the

Oregon border when winter set in. Evie and I boarded the horses at a large sheep farm in the mountains, promising to return for them in a few weeks, then the three of us caught a ride in a logging truck headed south. We each had a saddlebag worth of belongings, were as lean and fit as we'd ever be again, and had dreams enough to fuel the rest of our lives.

I returned to California, feeling like a grad student, ready to share the plant teachings I had received. It was 1972 and I was barely twenty-three. Everything in me felt green. I felt the plants singing in me, making me their song. A little corner of the natural foods store in the local town beckoned; I could open a small herb shop there. Community herbalism wasn't blossoming yet, but the time was ripe. And so my work began.

Almost overnight the little store thrived, not necessarily in financial ways, but the energy was green, full of life, vitality. People were immediately receptive to the plants; as if the community had been hungering for this green medicine. They came in for all kinds of illnesses and imbalances. From terrible things to common colds. Or they'd come in just to talk. They'd bring their children, sick friends, old folks. I was learning constantly. We opened a little teaching center at the top of the store, named it Stellar Energy Exchange, the *SEE* Center, and invited Ann Wigmore, Dr. Christopher, Dane Rudyard, and other such luminaries to teach there. I was blessed to meet those who carried the plant knowledge in the years when it had been suppressed, connecting a new generation to those that had gone before.

As I stand at this vantage looking back over the past thirty years since the doors to Rosemary's Garden first opened, I am struck by how quickly the seeds of herbalism rooted and blossomed. At the same time that the plants were calling me, they were calling hundreds of others as well. We would find each other and find that we spoke a common language as well. We discovered that we even had learned the same teachings about plants from the plants, sometimes word for word. We had gone to the same school and received similar teachings, though our classrooms were far apart. The plants were being heard. Their remarkable medicine honored again, their teachings listened to. Before our very eyes it seemed the Herbal Renaissance unfolded, the plants finding their place in America's life again. Later I started the California School of Herbal Studies, and later still moved to my home here on Sage Mountain in Vermont.

About this time I had another guiding dream, one that changed my entire relationship to my herbal work. I was visiting Equinox, a botanical sanctuary located in the northernmost range of the Appalachians, teaching a class with my close friend, Cindy Parker. This was my first visit to the area and I was, truly, in awe. More than 120 species of trees, 200 of fungi, and 600 of plants had been identified in the sanctuary so far. Many of them were favorite medicinals I'd used for years, but had only seen as photos in books. I was in heaven, side stepping over carpets of goldenseal, bloodroot, and hepatica as we explored the sanctuary. There was "old-growth" ginseng, large stands of black cohosh, and patches of mitchella blooming by the stream banks. My kind of class. We had hiked all day through this green maze, up and down hollers, over streams, through the woodlands. I was fully immersed, in awe of the green.

That night I laid my sleeping bag next to Golden Healing Pond and drifted off to sleep in a cacophony of night sounds one would expect to hear in a jungle. The forest here was that alive! The moon was full, beaming its luminous light directly on my bare head, drawing me into its wonder. Sometime during the night I heard my name called loudly enough that I woke from a light sleep. I heard it as clearly as one hears the ring of a bell on a clear dawn morning. It was the moon and, as I listened, she filled me with story.

Calling me back to the beginning of time, before time, before light. Into the stars, the void, the darkness, Cosmology unfolding. I saw the creation of light, of green matter, infinite creatures forming, exploding in tiny starlight patterns in the night. Millennia passing. How long did I dream?

Birthing from the most primal beginnings. Masses swirling, mountains pushing upward from massive plates shifting, Earth's surface upturned. Minerals and crystals forming, breaking up, forming into tiny particles of soil. Organic matter. Witnessing the first forms of life, algae, lichen, miniature ferns unfolding, tender, rooting in the tiniest substance of subsoil. All life as one, and I with it, witnessing. Though I saw it all, I remember little, and what I remember my limited vocabulary can't adequately describe. Unlike those poets who weave feelings into words that others can understand, and those brilliant minds that travel into time and come back with word paintings, I can share only that I lived this vision, understood, felt it grow in my being, and returned from the journey transformed.

> *Be not discouraged, keep on,*
> *there are divine things well envelop'd.*
> *I swear to you*
> *there are divine things more beautiful than words can tell.*
>
> —Walt Whitman

What I remember most of that night, and what impacted me long after, occurred toward the end of the journey. Earth was formed, darkness intermingled with light, and life forms were beginning to root. There were beings, bits of starlike energy that roamed this newly formed globe, creating and recreating the gardens of Earth. Some worked with rock, some with water, some (us among them) worked with the living green matter. Millennia passed. The gardens were in place. We are born again and again, renewed by the forces of life, each with our own spark. There are those who are called fully into the service of the green, unknowingly following a calling as ancient as life on Earth. These are the caretakers, keepers of the green, watching over the gardens through millennia. And throughout time, when the gardens are in danger, this energy is called to waken. The plants respond, animals respond, rocks and minerals respond, we respond.

PERHAPS THIS IS THE REAL WORK of herbalists these days. Not to argue politics, to create standards, or to set up regulatory boards for which we one day may be sorry, but rather to listen to the heartbeat of the plants, to seek to understand the medicine power that streams from it's source, to share that with others, and, ultimately, to restore the wild gardens, ensuring the continuing integrity of the plant communities.

Everywhere on this beautiful planet we witness the demise of nature, mostly by the erring hands of humans. The gardens of Earth are in danger. Perhaps this green renaissance is a call to action. What we know and love, we tend to care about. We get involved with. And we fight for. As we rediscover our relationship with plants—and what more intimate pathway than through the gateway of healing—it ignites a love, a passion for the green nations, and enables us to become caretakers of that which we love most. We are being called awake again in masses. Often the calling, at first, is not heeded or remembered, but somewhere deep inside a seed is buried, a recollection, a memory of the greening of the world. For many, it begins when we are children. Because children are closest to the Earth and hear best. And they understand the language of plants. As adults, we often enter into this relationship through our own pain, suffering, and injuries. We are touched by the plants, healed, and then become the healer.

This immersion happens many times. We are baptized, immersed fully in the world of the green, cared for, healed, taught, and renewed by the plants innumerable times. Having come full circle on this medicine journey, we stand poised. This is our give-back time, giving back to the plants that have given so much to us. The plants that have, once again, come to be our medicines.

While positive on one hand, this situation has engendered a unique set of challenges for wild medicinal plants and for the people who love and use them. Where are all those plants being used in our plant medicine coming from?

How are they thriving, these healing plants, in the intricate village system of their wild communities? How do the plant communities fare when important members of the medicinal clan disappear from their ecosystem, often harvested as ingredients for herbal products? What is known of the importance of ecological medicine, medicine of and for the Earth? After all, these are the powerful medicine plants, medicines as valued for the Earth's well-being and the health of the wild plant communities as for the two-leggeds, the humans, who have been dependent on them for thousands of years. What happens when the balance goes awry? When the medicine is removed from the community?

Is the ever diminishing population of these powerful medicine plants perhaps one of the reasons why there are so many more diseases attacking our native plant communities, as well as the human population?

INDIGENOUS PEOPLE HAVE LONG RECOGNIZED that all things in life are connected through a great web, and that disturbing one small plant from the ecosystem, from the great web of life, can cause the whole to go awry.

> *Frances Thompson, the English poet, once wrote:*
> *One cannot pluck a flower without troubling a star.*
> *If we cannot pluck a flower without troubling a star,*
> *what then if we lose a species?*
>
> —Loren Israelson

United Plant Savers was formed from that vision at Equinox. Our mission, simply stated, is to preserve and conserve our native medicinal plant resources, and the habitat they grow on, while providing sustainable renewable resources through organic cultivation. But the message is deeper. If we choose to use plants as our medicine, we then become accountable for the wild gardens, their health and their upkeep. We begin a cocreative partnership with the plants, giving back what we receive—health, nourishment, beauty, and protection. Plant by plant, sanctuary by sanctuary, acre by acre, we are regreening our Earth through the sacred awareness of the wild gardens. We need plants to survive; without them there is no life. And in some intricate way, which the plants understand better than we do, they need us as well. Our very humanness depends on them, and in return, they depend on us. To deny this is to deny our part in the great web of life. Their out-breath is our in-breath, our exhale, their inhale. One only understands this cocreative relationship as one enters the sacred territory, the cosmology, of the green.

> *The beauty and genius of a work of art*
> *May be reconceived,*

Though its first material expression be destroyed;
A vanished harmony may yet again
Inspire the composer;
But, when the last individual of a race of living beings
Breathes no more,
Another heaven and another earth must pass
Before such a one can be again.

—Belize City Zoo for Endangered Species

I CALL ON THE SPIRIT OF HERBS: BY JOHN SEED

I am tapping along keys on this laptop here in Siberia near the border of Mongolia, China, and Kazakhstan, where we join in the struggle of our Russian deep ecologist kin to preserve the Siberian Tiger (and the forests of Promorsky Krai on which they depend) and the Snow Leopard and the holy mountain Belukha and all her glaciers and the waters of the Katun River from damming.

I call upon the spirit of herbs to guide these fingers.

I call upon the migrating molecules presently in these keys and upon the electrons that dance across the backlit screen to join in embrace with consciousness in an unfolding of Truth that may awaken awe in the most calloused human heart.

I call upon the spirit of herbs to speak as they have spoken through shaman upon shaman buried in the layer rock of consciousness, I call on them to speak of deep ecology and of the closing decade of this millennium where we face a challenge perhaps as great as the introduction of oxygen into our ancient methane world. While watching the rate of extinctions soar exponential I remember the Cretaceous when our ancestors were among the mere 5 percent of species that made it through the momentous events of that epoch, a strong record of success for ourselves and the matrix in which we are embedded, in which we dance.

. . . . and I was the stars,
Boiling with light, wandering alone, each one the lord of his own summit; and I was the darkness
Outside the stars, I included them.
They were part of me.
I was mankind also, a moving lichen
On the cheek of the round stone . . . they have not made words for it, to go beyond things, beyond hours and ages,
And be all things in all time,
in their returns and passages
in the motionless and timeless centre,
In the white of the fire . . . how can I express the excellence I have found . . .

—Robinson Jeffers, *Not Man Apart*

In 1979, after years of meditation, of growing organic food in community, of planting seeds that grew from the seeds I had planted the season before and feeling body and mind grow awestruck from the wonder of it all, I found myself somehow embroiled in the defense of a little forest.

What would the world be, once bereft
Of wet and of wilderness? Let them be left,
O let them be left, wildness and wet;
Long live the weeds and the wilderness yet

—Gerard Manley Hopkins, "Inversnaid"

I stood with my fellows on that bewildering line between the human world and the dark green vegetation, but this time not with an eye to conquest, to the "development" of this forest, but standing with the trees, alongside the trees to prevent their destruction.

I speak for the trees as the trees have no tongues
And I'm telling you sir AT THE TOP OF MY LUNGS . . .

—Dr. Seuss, *The Lorax*

Out of this small beginning came the Rainforest Information Centre, The Council of All Beings workshops, and my personal path of speaking for Earth. How little could I have known what strange paths I would walk once I remembered the ancient language of Earth and plant and animal and gave myself up to its voice.

Our faith imposes on us a right and duty
to throw ourselves into the things of the earth.

—Teilhard de Chardin

The Council of All Beings is the ceremonies and rituals that bring our deep ecology to life. Here we have a practice where we approach a leaf as though approaching our revered zen master. We breathe to this leaf the oxidized carbon of our body. We do so with the gratitude and the generosity that is the signature, the clue to the Nature of which we are a fragment.

As we add consciousness to the ancient processes of sharing respiration, we savor the leaf in our imagination. Now we must notice and then lay aside our prejudice that humans are the only ones capable of consciousness in this transaction, this holy communion that accompanies our every breath. We consciously nourish a leaf and invite the leaf to nourish us, not just with the oxygen it creates, but with further communications.

The most "primitive" peoples, living deeply embedded in their "environment," all practice ceremonies and rituals that affirm and nourish the interconnectedness, the interbeing, of the human tribe with the rest of the Earth family. This would indicate that the propensity to lose this connection is not just a modern phenomenon but is rooted deep within our humanity. We moderns, however, in our arrogance and "enlightenment," have ridiculed such practices, attempted to assign them to the realm of superstition. Ritual has become "empty ritual." Thus our connections are in tatters and the world torn asunder. Having ridiculed such rituals, we did not participate in them; not participating in them, we lost our place in the world. And now, how are we to recover our ecological self? Mere ecological ideas, no matter how deep, cannot save us.

It is tempting to see the Council of All Beings as a therapy. However, a therapy surely is something that ends once the patient is cured. And this disconnectedness from the living Earth is not something that leaves us, once "cured" by our workshops, free to go back to modern ways. The Hopi, for example, living in the most ancient continually inhabited villages in the Western Hemisphere, still don their animal masks and re-member humanity to the body of the Earth.

The Council of All Beings and other re-Earthing rituals must be reincorporated into the very foundations of society or else no amount of "conservation" of nature, no "saving" of "representative areas" in national parks, will stop the hemorrhage whence the very lifeblood drains from the Earth.

Plants! Herbs!

Of course they feed us! Of course they heal us! Of course they get us high! We've been co-operating together for eons before arrogance and amnesia set in, they are the manner in which we are rooted in biology, they mediate between us and the sun.

They lay between us and the inorganic world that they suck spectacularly into biology as we suck them in turn into the peculiar consciousness that now reflects back upon these things.

Much information and wisdom has been methodically suppressed. From the burning of over 100,000 European herbalists grew the modern masculinist medical profession. How are we to reclaim our selves? Can this be learned by the modern mind in the sense of "adding something on" to our existing knowledge base? Another acquisition for the mind?

To reclaim ourselves we must tap once again that ancient participation in the deep life of the Earth of which we are a part. We must know that world as surely as we think we know the one in which we have lived so long. To know it we must use some faculty other than mind. A beginning of this is imagination.

Imagination itself, of course, is a recent extrusion of the animal through some of its more recent exuberant bulges of forebrain and is embedded inextricably in the organic matrix from which it blooms. So, I invite my imagination (and yours) to trace a path back. Beyond the point where imagination is rooted in this physical body (this incredible record of success upon success, 4 billion years of organic symbiosis). Back to where the animal is rooted in the vegetable: And through that to wedding our mutual and complementary roles in the sharing of breath; through the fact that primary production on this Earth is vegetable and the animal world must first eat the vegetable world to exist, and only later can animals feed on other animals; through the miracle of the brilliant stroke of creativity when plants first decided to spring forth and capture sunlight, to eat the photons that fall upon this barren rock. Ahhh, photosynthesis!

Back to the ultimate miracle of sunlight itself, energy with the propensity, the delight, to weave itself into all this.

Deep ecology concerns the interconnectedness of all things, the way that all beings—plant, animal, human beings—are part of a larger organism, arcs in vast circuits. We are all cells in the vast body of Earth as Earth itself participates in wider solar and galactic reciprocities.

Inside this clay jug are canyons and pine mountains
And the maker of canyons and pine mountains
All seven oceans are inside
And hundreds of millions of stars

—Kabir, *The Kabir Book* (versions by Robert Bly)

The gene pool: the ancient partnership between the herbs who capture sunlight and make it all possible, photosynthesis, creating the basis for animal and human life who can't transform the sun directly but must rest upon the layer beneath, just as the vegetation rests upon geology, and the animal layer rests upon the vegetable.

Our evolutionary journey.

In the fourth "World Rainforest Report," Queensland zoologist Peter Dwyer noted that the New Guinea highlanders find the rain forest wildlife not only good to eat, but also "good to think." He goes on to say:

> Whilst we don't eat our rainforests, we do become enmeshed in our perceptions and thinking about them until they suddenly and

> vividly possess for us values that we can only identify as symbolic, intrinsic and—with some desperation—as spiritual. The tropical rainforests are primitive and ancient ecological systems whose origins stretch backwards through the emergence of the flowering plants in Jurassic times over 135 million years ago to the plants preserved in the coal measures of the carboniferous millions of years before that and which appear to us today in the form of plastics. Such is biogeochemical continuity.

Dwyer's ability to see rain forests of hundreds of millions of years ago embedded in the plastics of the present age is a good example of the *psychological* effects of rain forests upon people who spend their time in them. *Psycho*biogeochemical continuity?

For psyche grows from a foundation of animal and vegetable. The personality conditioned in the modern human has found it convenient to filter out conscious awareness of this process. Let us now invite ourselves to dismantle these filters and begin to extend our awareness and identity to the points where our intersections with animal and vegetable take place.

For now our glorious and crucial task is to take this magnificent heritage of our evolutionary journey, remember it, recapitulate it, call on its power and momentum to aid us on a further leap into creation. Wedded to our friends the herbs, we prepare for another leap into complexity, into articulation.

> *Good people*
> *most royal greening verdancy,*
> *rooted in the sun,*
> *you shine with radiant light.*
>
> —Hildegard of Bingen

In Australia, two sisters by name of Bradley came up with an exciting technique by which we may slowly invite back the original biotic community from denuded and scarred landscapes.

Their method is essentially simple, unheroic. Nothing needs to be planted. All we have to do is to learn to distinguish the exotic from the native species right from when the tiny seedlings first emerge. Then we remove the exotics without treading on the natives.

This must be done painstakingly and methodically. And we must start from the strongest expression of native vigor in our management area. That is, let's say there's an erosion gully breaking our hearts, and we want to start the repairs there at once but any attempt to heal it is doomed. What we must first do is discover and begin from some least-damaged spot. For example, in a park of introduced grasses, we find a corner that the mower couldn't reach where a few annual weeds flourish. We begin from this corner and move out. Or perhaps among all the annuals, we spot a few pioneers and begin carefully clearing the exotics from around them.

Slowly at first, the pioneer natives emerge, creating shade, soil, microclimate, then the next stage in the succession, till finally, years into the process, the conditions are ripe for the reemergence of the climax species. We will never know whether the seeds were introduced by a bird flying by or perhaps lay dormant beneath the soil waiting for conditions that allow their germination. But years into the Bradley Method of an area, to our amazement, climax native trees begin to emerge, perhaps unseen for generations and unremembered.

With each succeeding season the process becomes more vigorous and robust. By the time the spreading cover of burgeoning native vegetation reaches the erosion gully, it takes it in its stride.

The Bradley Method is more than a handy tool that teaches us how to repair the simplification that we have wrought. It also provides us with an extremely useful metaphor for reclaiming our own native wisdom. Buried under millennia of conditioning, suppressed by inquisitions, ridicule, and doubt lie unimagined potentialities to reinhabit our interbeing with the world, to fully participate in the world once again.

It's no use to learn something new to accomplish this transformation, however. Merely to discard the false certainties of this age (of our "triumph" over nature), to weed out our anthropocentrism, is to create the conditions for healthy psyche to spring forth anew. We must learn only to discern and discard those exotic ideas that suppress our old ways of seeing to allow ancient knowing its place in the sun. Of course, we may wish to begin at once with the transformation of the CEO of some scumbag corporation. However, preaching to the choir is the way to go, strengthening and nurturing deep ecology wherever we see it emerge rather than attempting to transplant it into souls lacking the necessary nutrients and microclimate.

In the Council of All Beings we practice in the ways of old, the generosity to lend our voice, to lend an ear, to spare a sigh, to hear the tale of the many nonhuman beings of Earth. Thus consciousness completes the cycle and ensures once again and again and again that this frontal lobe does not succeed in its fevered excitement to leap forward and out of matter altogether. We practice rooting ourselves in the wisdom of plant, allowing consciousness again and again to travel along, to open up the pathway to our source—through animal to plant, through plant to geology and sunlight, inviting the sunlight in the plants that fuels these fingers to dictate:

> O humans, are you truly tired of our dance together? Have we not served each other well through the endlessness of our journey? Remember humans, remember.
>
> Remember the vast abyss of time and that all of the poor certainties of your ideologies and religions are merely an expression of the last split second.
>
> Remember humans, that you too are leaves on the tree of life. The sap of the tree pulses through your veins. Of course you can allow your being to glide down the stream. Invite yourself to slip back into and through the tree, into each other leaf and into the ground of being itself. All you need do is remember and slide juicy back into the flow.

The force that through the green fuse drives the flower
Drives my green age; . . .
The force that drives the water through the rocks
Drives my blood

—Dylan Thomas, "The Force that through the Green Fuse Drives the Flower"

In service to the larger pictures, deep ecologists surrender the momentary dreams of this recent time to offer our lives, our days, our minds, to the service of plants and the animals and the species and communities and ecosystems and life-support systems. Moment to moment, ritual by ritual, action by action, we surrender tiny human desires to the great lust of the universe to fashion itself ever more complex and subtle. Meaning and order emerge spontaneously. No need to strive, no need to build: Community emerges "by itself" among those who thus surrender to the same larger picture. Like iron filings consenting to the stroke of the same ancient magnet, we fall easily into harmony and our actions may approach once again the divine greatness of the rest of the universe.

O sweet spontaneous
earth how often have
the
doting

fingers of
prurient philosophers pinched
and
poked

thee

—ee cummings, "O sweet spontaneous earth"

For eternity, without the mixed blessing of thinking, we spontaneously, instinctively, reflexively danced in harmony with the symphony of Earth. Just think of it! Your ancestors, my ancestors, without thought for 4 billion years succeeded, succeeded! Each generation our ancestors managed to breed before being consumed, while at each step of the journey thousands, millions fell gladly by the wayside, spent, donating their flesh, the sunlight they had accumulated and stored, to those few at each step who carry the living flame forward.

To dwell upon such things is to allow the humility that blooms effortlessly in this medium. To allow the possibility of unmade synthesis and co-operation and mutual aid with countless beings whose surrender needs no thought is the beginning. To do this regularly is to join the shamans and healers of all times, the ceremonies of all natural peoples who thus remained in harmony with their world. They did not expel themselves as we have done into the cold illusions of separateness that may be the chief characteristic of our kind.

To surrender thus is to fall, not back, but forward into an age where the knowledge that we have gained (at what cost!) may once again be plowed back into the eternally fertile soil of this Earthly existence.

We make the path by walking:
You, walker, there are no roads
Only wind trails on the sea

—Antonio Machado, "I Never Wanted Fame"

EPILOGUE

IN MY TURN I HAVE WALKED WITH MY SON IN the deep forest. I have lain next to him and felt something leave my body and enter his. He needs it less often the older he becomes. Still, sometimes he is unsettled and paces the floor and a peculiar look comes over his face. Agitated, he will ask if he can lie next to me. And in silence I hold him and something in him opens up and a flood flows out of me into him. The color and tone of his skin changes and his breathing slows and deepens and eventually he sighs and is filled once more. And I know that in his time he will pass this on as it was in turn passed into me. And perhaps also, one day, he will bend over and cup his hands, and ask his child or grandchild or some child:

"Here, have you ever tasted this water?"

Notes

Chapter One

1. Stephen Harrod Buhner, *The Taste of Wild Water* (Coeur d'Alene, Id.: VisionWeaving [now Raven] Press, 2000). Slightly revised for this book.

Chapter Three

1. Melvin Gilmore (1977) quoted in Kelly Kindscher, *Medicinal Wild Plants of the Prairie* (Lawrence: University of Kansas Press, 1992), 51.
2. Stevens (1961) quoted in ibid., 51.
3. David Ehrenfeld, *The Arrogance of Humanism* (New York: Oxford University Press, 1978), 8.
4. Rupert Sheldrake, *The Rebirth of Nature: The Greening of Science and God* (New York: Bantam, 1991).
5. Ibid., 43.
6. Ibid., 42; see also Rupert Sheldrake, *Morphic Resonance and the Presence of the Past* (Rochester, Vt.: Park Street Press, 1995); Richard Grossinger, *Planet Medicine* (Boulder, Colo.: Shambhala, 1982); and Alston Chase, "Harvard and the making of the Unibomber" *Atlantic Monthly* (June 2000).
7. Ehrenfeld, *The Arrogance of Humanism,* 10.
8. Sheldrake, *The Rebirth of Nature,* 53.
9. Ibid.
10. James Lovelock, *Healing Gaia: Practical Medicine for the Planet* (New York: Harmony Books, 1991), 38.
11. This kind of experimentation has continued with black people, for example the Tuskeegee study in which black men with syphilis were not told of their condition and were left untreated for decades. Researchers today are pursuing similar approaches with Africans infected with AIDS, see Jerome Groopman, "In an AIDS Study, the devil is in the details" *New York Times,* April 2, 2000, 4.
12. Richard Milton, *Shattering the Myths of Darwinism* (Rochester, Vt.: Park Street Press, 1997), 186.
13. Raoul Hilberg, speaking at the Conference on Evil, recorded by Bill Moyers for the NPR program "Facing the Evil," 1988.
14. quoted in Stephen Harrod Buhner, *One Spirit, Many Peoples* (Boulder, Colo.: Roberts Rinehart, 1997), 181.
15. Douglas Martin, "Keith Mant Dies at 81: Pathologist helped convict Nazis," *New York Times,* November 26, 2000, 54. The Soviet biological weapons program was, in many ways, not so different, though it was directed at

Americans rather than Jews or Slavs. The latest information reveals that physicians and scientists used their knowledge to create highly virulent and resistant forms of diseases such as smallpox, anthrax, tularemia, and plague that could not be treated with current medical knowledge. The director of that program was Kanatjan "Ken Alibek" Kalibekov, who defected to the United States after the breakup of the Soviet Union. "Like every doctor," the *New York Times* notes, "he took an oath to preserve life and guard against medical abuse. But after doing so, he devoted himself for 17 years not to healing but the science of germ warfare—devising new ways of harnessing deadly microbes to maim and kill people." (William Broad, "Finding penance in going public about making germs into bullets," *New York Times,* June 27, 1999, 7.) He was motivated, Alibek says, by careerism, by the privileges that came with it "like personal cars, special telephone lines, special medical treatments—you know, a lot of such things." And, of course, patriotism. "The moral argument for using any available weapon against an enemy threatening us with certain annihilation seemed to me irrefutable" (ibid; Ken Alibek, *Biohazard: The Chilling True Story of the Largest Covert Biological Weapons Program in the World Told From Inside By the Man Who Ran It* (New York: Random House, 1999). After the collapse of the Soviet Union many of the scientists and physicians sold the altered germs to other countries and splinter groups for use as weapons to, as Alibek says, "support themselves" (ibid; the ethical principles of science and medicine are often sacrificed this way, e.g., see Lord Phillips, June Bridgeman, Malcom Ferguson-Smith, *The BSE Inquiry: vols I–XVI* (London: The Stationary Office, 2000).

16. Christopher Stone, *Should Trees Have Standing?: Toward Legal Rights for Natural Objects* (Los Altos, Calif.: William Kaufmann, 1974), 7.
17. Ibid.
18. Theodore Roszak, "Eco-Psychology," *Earth Ethics* 5, no. 2 (winter 1994); see also David Orr, "2020: A proposal," *Conservation Biology* 14, no. 2, (April 2000): 338–41 for interesting comparisons between justifications for continued technological dominance of Earth and the necessity of slavery. In support of the necessity of continued technological dominance of Earth, Ken Wilber and others insist that the attribution of equal value by people to the other members of the physical world will result in a paralyzing of pragmatic action. However, an examination of other cultures, such as the Jains in India, that have extended such rights shows that pragmatic action is not paralyzed.
19. Quoted in Stephen Harrod Buhner, *Sacred and Herbal Healing Beers* (Boulder, Colo.: Siris Press, 1998), 16.
20. Ken Wilber, *A Brief History of Everything* (Boston: Shambhala, 1996).
21. Mary Oliver, *House of Light* (Boston: Beacon Press, 1990).

Chapter Four

1. Tom Wakeford, "Undressing the evolutionary emperor," *TREE* 13, no. 5, (May 1998): 211; see also: Richard Milton, *Shattering the Myths of Darwinism* (Rochester, Vt.: Park Street Press, 1997); Egbert Giles Leigh, "The modern synthesis, Ronald Fisher and creationism," *TREE* 14, no. 12 (December 1999): 495; Eva Jablonka, Marion Lamb, Eytan Avital "'Lamarckian' mechanisms in Darwinian evolution," *TREE* 13, no. 5 (May 1998): 206.

2. Lynn Margulis, *Symbiotic Planet: A New Look at Evolution* (New York: Basic Books, 1998); Lynn Margulis and René Fester, *Symbiosis as a Source of Evolutionary Innovation: Speciation and Morphogenesis* (Cambridge: MIT Press, 1991); Lynn Margulis and Lorraine Olendzenski, *Environmental Evolution: Effects of the Origin and Evolution of Life on Planet Earth* (Cambridge: MIT Press, 1992); Lynn Margulis and Dorion Sagan, *What Is Life?* (New York: Simon and Schuster, 1995); Lynn Margulis, Karlene Schwartz, Michael Dolan, *Diversity of Life: The Illustrated Guide to the Five Kingdoms* (Sudbury, Mass.: Jones and Bartlett, 1999).
3. Margulis, *Symbiotic Planet*, 48–49.
4. Edward O. Wilson, "Biophilia and the conservation ethic," *Biophilia Hypothesis*, ed. Edward O. Wilson and Stephen Kellert (Washington D.C.: Island Press, 1993), 39.
5. Masanobu Fukuoka, *The Road Back to Nature* (New York: Japan Publications, 1987) 49, 50.
6. James Lovelock, *Healing Gaia: Practical Medicine for the Planet* (New York: Harmony Books, 1991), 17.
7. Reed Noss, "The naturalists are dying off," *Conservation Biology* 10, no. 1, (1996): 1.
8. Mircea Eliade, *The Forbidden Forest* (Notre Dame, Ind.: University of Notre Dame, 1978), vii–viii.
9. Ibid., vii.
10. John Gardner, *The Art of Fiction: Notes on Craft for Young Writers* (New York: Vintage Books, 1991), 30–31.
11. Ibid., 31.
12. Ibid., 36–37.
13. Ibid., 46.
14. See, for instance: David Orr, "Virtual nature," *Conservation Biology* 10, no. 1 (February 1996): 8–9.
15. James Hillman, *The Soul's Code* (New York: Random House, 1996), 84.
16. See, for example: Richard Rhodes, "The media-violence myth" *Rolling Stone* (November 23, 2000), 55–58.
17. Stephen Kellert, "The biological basis for human values of nature," in *Biophilia Hypothesis*, 42–43.
18. Gary Paul Nabhan and Sara St. Antoine, "The loss of floral and faunal story: The extinction of experience," in ibid.
19. Gardner, *The Art of Fiction*, 43.
20. Quoted in Barbara Griggs, *Green Pharmacy* (Rochester, Vt.: Healing Arts Press, 1991), 266.
21. Quoted in ibid., 261.

Chapter Five

1. Christian Daughton and Thomas Ternes, "Pharmaceuticals and personal care products in the environment: Agents of subtle change?" *Environmental Health Perspectives* 107, Supplement 6 (December 1999).
2. Sheryl Gay Stolberg, "A drug plan sounds great, but who gets to set prices?" *New York Times,* July, 9, 2000, section 4, 1.
3. Andrew Sullivan, "Pro pharma," *New York Times Magazine,* October 29, 2000,

24; Kathy O'Connell, "Pill poppin' nation," *The Inlander,* July 22, 1998; Milt Freudenheim, "Consumers across the nation are facing sharp increases in health care costs in 2001," *New York Times,* December 10, 2000, section 1, 40; Sonya Ross, Associated Press "Clinton: GOP drug plant is 'baloney,'" July 31, 2000—on America On Line (AOL); Eli Ginzberg and Panos Minogiannis, "Medical care in the U.S.—Who is paying for it?" *Journal of Practice Management* 15, no. 5 (2000); "Because of the costs many poor Americans and most people in poorer nations are simply out of luck," Donald McNeil, "Do the poor have a right to cheap medicine?" *New York Times,* June 25, 2000, 18.

4. Peter Montague, "Headlines: Paydirt from the human genome," *Rachel's Environment and Health Weekly* #702, July 6, 2000, www.rachel.org (hereafter *Rachel's*).
5. Quoted in ibid.
6. Daughton and Ternes, "Pharmaceuticals and personal care products in the environment."
7. Rx List: The Internet Drug Index (February 24, 2000), www.rxlist.com.
8. Peter Montague, "Excrement happens," part 1, *Rachel's* #644, April 1, 1999; Peter Montague, "Excrement Happens," part 2, ibid., #645; Abby Rockefeller, "Civilization and sludge: Notes on the history of the management of human excreta," *Current World Leaders* 39, no. 6 (December 1996): 99–113.
9. Montague, "Excrement happens," part 1.
10. Federal Register, U.S. EPA November 9, 1990, quoted in ibid.
11. Jerry Schwartz, "Philadelphia's well-traveled trash," Associated Press, AOL news online, September 3, 2000.
12. Montague, "Excrement happens," parts 1 and 2.
13. Greenpeace Toxic Site Online; www.greenpeace.org.
14. Greenpeace—Stats, www.greenpeace.org/~toxics.
15. Daughton and Ternes, "Pharmaceuticals and personal care products," 27; Heberer, "Determination of the clofibric acid and N-(phenylsulfonyl)-sarcosine in sewage, river and drinking water," *Journal of Environmental Analytical Chemistry* 3 (1997): 1; T. Heberer, et al., "Detection of drugs and drug metabolites in ground water samples of a drinking water treatment plant," *Fresenius Environmental Bulletin* 6 (1997): 438.
16. Ibid.; B. Hallung-Serensen, et al., "Occurrence, fate and effects of pharmaceutical substances in the environment—a review," *Chemosphere* 36 (January): 357; Janet Raloff, "Drugged waters," *Science News* 152 (March 21): 187.
17. Ibid.
18. Rx List: The Internet Drug Index.
19. Sandeep Jauhar, "More fans for drugs that fight cholesterol," Science Health Online, September 16, 2000
20. Peter Montague, "Fish sex hormones," *Rachel's* #545, May 8, 1997; Daughton and Ternes, "Pharmaceuticals and personal care products."
21. Janet Raloff, "Excreted drugs: something looks fishy" *Science News* 157, no. 25 (June 17, 2000); Montague, "Fish sex hormones," *Rachel's;* Montague, "Drugs in the water" *Rachel's* #614, September 3, 1998.
22. Ric Jensen, Online posting, February 14, 2000, Water Conference in

Minnesota: Conference outline, "Endocrine disrupting chemicals in water," www.twri.tamn.edu; see also Peter Montague, "Chemicals and the brain," parts 1 and 2, *Rachel's* #s 499 and 501, June 20 and July 4, 1996.

23. Montague, ibid.; see also Peter Montague, "Infectious disease and pollution," *Rachel's* #528, January 9, 1997.
24. Anita Manning, "Menopause relief may be a gamble," *USA Today,* August 3, 1998.
25. Daughton and Ternes, "Pharmaceuticals and personal care products," 26.
26. Peter Montague "Girls are reaching puberty early," *Rachel's* #566, October 2, 1997.
27. Lisa Belkin, "The making of an 8-year-old woman," *New York Times Magazine,* December 24, 2000, 38–43.
28. Janet Raloff, "More ground, lake, and stream waters test positive for drugs," *Science News,* 157, no. 14 (April 1, 2000).
29. Daughton and Ternes, "Pharmaceuticals and personal care products," 22.
30. Ibid.
31. Ibid.
32. Ibid.
33. Ibid.
34. Ibid., 35.
35. Chandra Twary, "Premature sexual development in children following the use of placenta and/or estrogen containing hair product(s)," *Journal of Pediatric Research* 135 (1994): 108A.
36. Kenny Ausubel, *When Healing Becomes a Crime* (Rochester, Vt.: Healing Arts Press, 2000), 240, 232.
37. Ibid., 237.
38. Ibid.
39. P.A. White and J.B. Rasmussen, "The genotoxic hazards of domestic wastes in surface waters," *Mutat Res,* 410 (1998): 223–36.
40. Daughton and Ternes, "Pharmaceuticals and personal care products," 31.
41. Ibid.
42. Nicholas Lenssen, *Nuclear Waste: The Problem That Won't Go Away,* Worldwatch paper 106 (Washington, D.C.: Worldwatch, December 1991).
43. U.S. Council for Energy Awareness, "What do you know about low-level radioactive waste?" http://starfire.ne.uiuc.edu, September 17, 2000; "Low-level radiation fact sheets," www.ag.ohio-state.edu/~rer/.
44. Ibid.; "Industry and Earth Sciences 1996 Annual Report," www.iaea.org; "Isotopic irradiation," *McGraw-Hill Encyclopedia of Science and Technology,* 7th edition, 1992, 460; Patrick McMahon, "Cancer survivors support reactor," *USA Today* October 23, 2000, 3A.
45. McGraw-Hill, "Isotopic irradiation," ibid., 460; Anita Manning, "Irradiation options packaged to produce pathogen-free meat," *USA Today* May 10, 1999, 8D.
46. Patricia Boiko, "Physicians for social responsibility against the FFTF" (Seattle: Washington Physicians for Social Responsibility Action Paper, 1999).
47. Ibid.

48. "Low-level radiation fact sheets," www.ag.ohio-state.edu/~rer/; McMahon, "Cancer survivors support reactor"; Lenssen, *Nuclear Waste;* Peter Montague, "Low-level radioactive waste—part 1: fifty years of failure" *Rachel's* #302, September 9, 1992; Peter Montague, "Radiation—part 2: bad news about low-level radiation" *Rachel's* #184, June 6, 1990; Peter Montague, "Landfilling low-level radioactive is a problem for all states," *Rachel's* #69, March 21, 1988; Peter Montague, "Dangers of radiation are now increasing in municipal waste," *Rachel's* #62, February 1, 1988.
49. Robert Soderstrom, "Testimony of Robert Soderstrom, M.D." April 7, 1992, www.greenlink.org.
50. Eggermont, GX, et al., "Decay management of nuclear waste in university and hospital," 1997, www.inel.gov/resources/ research/.llrw/1997Conference; Soderstrom, "Testimony"; McGraw-Hill, "Isotopic irradiation," 460.
51. Lenssen, *Nuclear Waste;* John Gofman, *Radiation From Medical Procedures in the Pathogenesis of Cancer and Ischemic Heart Disease* (San Francisco: Committee for Nuclear Responsibility, 1999).
52. Stephen Buhner, "Ecological medicine," *The Healer's Review* 2, nos. 2 and 3 (double issue) (Boulder, Colo.: Colorado Association of Holistic Healing Professionals, July 1991).
53. Eggermont, GX, et al., "Decay management of nuclear waste in university and hospital."
54. Soderstrom, "Testimony."
55. United States Nuclear Regulatory Commission, Office of Nuclear Material Safety and Safeguards, December 21, 1999. NRC Information Notice 99–33. "Management of Wastes Contaminated with Radioactive Materials."
56. Quansi Salako and Sally Denardo, "Analysis of long-lived radionuclidic impurities in short-lived radiopharmaceutical waste using gamma spectrometry," *Health Physics* 72, no.1 (1997): 56.
57. Cemetery/mortuary/crematorium pollution: There are approximately two and one-half million deaths each year in the United States and many of the dead's bodies contain substantial quantities of pharmaceuticals. This is especially true of the very old or very ill, all of whom are usually given exceptionally large doses of pharmaceuticals, chemodrugs, and/or radiation prior to death. All of these substances infuse bodily tissues and fluids. During embalming, embalming chemicals are injected through an artery while bodily fluids are drained from another. The contents of the abdominal cavity are removed as well, usually as liquids through powerful suction procedures. In many instances all the fluids are simply drained into the local municipal sewer system and become part of the local waste treatment stream. A few mortuaries in some states treat some decedents as medical waste and transfer either liquids or the bodies themselves to waste containers shipped to central medical waste treatment locations. (Alan Bavley, "Babies die, then more tragedy. A funeral home apparently handled the babies as medical waste instead of cremating," *Kansas City Star,* July 6, 1999.)

    Approximately half of all people are buried, and although embalming slows the reintegration of the body into the ecosystem it eventually does occur. Though no figures are available on the leaching of pharmaceuticals from these sources, early research in Australia suggests treating cemeteries as specialized landfills because of leachate concerns. Testing of ground water in Australia shows leaching of identifiable substances from graves into ground water despite cemetery design. Because of the centralized gathering of bodies in one place and such newer practices as using single graves for multiple burials over time at differing depths, cemeteries produce consistent ground water contaminant plumes that contain such things as infectious bacteria, heavy metals from mercury in dental fillings, and pharmaceuticals. New grave sites (because of loosening of the soil) and even the coffins themselves add to the problem by acting as water accumulators ("bucket and sponge"), which serves to concentrate leachates in ground water. (Boyd Dent and Michael Knight, "Cemeteries: A Special Kind of Landfill," National Centre of Groundwater Management, University of Technology, Sydney, Broadway, NSW 2007 Australia, n.d. [c. 1997].)

About one and a quarter million people are cremated each year, usually after embalming. (Sidun Funeral Group, "Methods for Disposition of the Decedent," www.sidun.com.) The embalming chemicals themselves are exceptionally toxic. OSHA (Occupational Safety and Health Administration) regulations require that embalmers wear full-body coverings and respirators while embalming. Until recently crematoriums, like hospitals, were exempt from federal air-quality regulations. In the past decade concern has risen about the levels of this kind of pollution, especially from neighbors objecting to smokestack emissions. (Sharyn Obsatz, "Crematorium spooks neighbors," *The Detroit News*, May 7, 2000; Don Stacom, "Leete-Sevens seeks to end opposition to crematorium," *The Hartford Courant,* June 5, 1998; Charlie Able, "What's in that thick smelly smoke? Don't ask, just stay indoors," *Seattle Post-Intelligencer,* November 7, 2000; Funeral Consumers Alliance, "What You Should Know About Embalming," www.funerals.org.) Environmental impact assessments have noted that the combination of embalming chemicals, dental fillings, pharmaceuticals, and incineration at crematoria produces significant quantities of mercury and dioxin pollution. Small crematoria are calculated to produce three kilograms of cumulative loading of mercury to surrounding land per year. (Douglas Cross, "Environmental issues of the proposed Allerton Crematorium," Dartingon, South Devon, September 4, 1996. www.doublef.co.uk/texts/ecol/crem.htm.) Pharmaceuticals are recognized as a unique difficulty not faced by older generations of embalmers. Making the bodies look natural has become more of a problem. (Ausubel, *When Healing Becomes a Crime*, 239–40).

58. Environmental Protection Agency, "EPA Tracking Program for Medical Waste Starts Today," www.epa.gov.
59. Health Care Without Harm, *Greening Hospitals: An Analysis of Pollution Prevention in America's Top Hospitals* (Falls Church, Va.: Center for Environment and Justice, 1998); "EPA tracking program;" *Infectious Wastes News* 13, nos. 1–24 (1998), www.worldwastes.com.
60. Health Care Without Harm, *Greening Hospitals,* 10.
61. Ibid.; see also, Michael Malloy, "EPA set to issue incinerator rule August 15, enviros say they'll sue for stricter standard," *Infectious Wastes News,* July 15, 1997.

62. Ibid.
63. Ibid.
64. Hollie Shaner, "Back to glass? No thank-you to phthalates," *Vermont Registered Nurse* 64, no. 2, 6 (undated photocopy).
65. Joel Tickner, et al. *The Use of Di-2Ethylhexyl Phthalate in PVC Medical Devices: Exposure, Toxicity, and Alternatives* (Lowell, Mass.: Lowell Center for Sustainable Production, 1999); Health Care Without Harm Health Alert, "Vinyl IV bags leach toxic chemicals," Falls Church, Va., n.d.; Greenpeace, "Patients exposed to toxic chemicals from PVC medical products," Greenpeace online, www.greenpeace.org.
66. Ibid.
67. Tickner, et al., ibid.
68. Health Care Without Harm, *Greening Hospitals,* 18.
69. Health Care Without Harm, *Mercury Thermometers and Your Family's Health* (Falls Church, Va.: Center for Health, Environment and Justice, n.d.).

Chapter Six

1. W.J. Powell, "Molecular Mechanisms of Antimicrobial Resistance," Technical Report no. 14, February 2000, www.plant.uoguelph.ca/safefood/ab-res-ppr-wjp.htm, p. 2.
2. Gene Emery, "Study: Antibiotics Resistance Common, Rising," Reuters, AOL-online, December 30, 2000.
3. Jan Dirk van Elsas and Kornelia Smalla, "Antibiotic (kanamycin and streptomycin) resistance traits in the environment," *Proceedings of the Workshop on Key Biosafety Aspects of Genetically Modified Organisms,* Braunschweig, Germany, April 10–11, 1995, www.bba.de/gentech/workshop.htm.
4. Christian Daughton and Thomas Ternes, "Pharmaceuticals and personal care products in the environment: Agents of subtle change?" *Environmental Health Perspectives* 107, Supplement 6 (December 1999); Stuart Levy, *The Antibiotic Paradox* (New York: Plenum Press, 1992); Peter Montague, "Hidden costs of animal factories," *Rachel's* #690, March 9, 2000; Janet Raloff, "More ground, lake, and stream waters test positive for drugs," *Science News Online,* April 8, 2000, www.sciencenews.org.
5. Elsas and Smalla, "Antibiotic resistance traits in the environment."
6. Levy, *The Antibiotic Paradox,* 94.
7. Ibid., 75.
8. Ibid., 3.
9. Ibid.
10. Quoted in Barbara Griggs, *Green Pharmacy,* 261.
11. R.L. Berkelman and J.M. Hughes, "The conquest of infectious diseases: Who are we kidding?" *Annals of Internal Medicine* 119, no. 5 (1993): 426–27.
12. Marc Lappé, *When Antibiotics Fail,* 187.
13. Powell, "Molecular mechanisms of antimicrobial resistance."
14. Lappé, *When Antibiotics Fail,* xvi.
15. Elsas and Smalla, "Antibiotic resistance traits in the environment," 24.
16. Powell, "Molecular mechanisms of antimicrobial resistance," 8.
17. Levy, *The Antibiotic Paradox,* 101.

18. Quoted in Buhner, *Herbal Antibiotics*, 9.
19. Ibid.
20. Quoted in Powell, "Molecular mechanisms of antimicrobial resistance," 12.
21. Catherine Amst and Kerry Capell, "Tuberculosis roars back," *Business Week*, October 2, 2000, online version.
22. Lisa Belkin, "A brutal cure," *New York Times Magazine,* May 30, 1999, 34.
23. Associated Press, "Gonorrhea's resistance grows," AOL-online, September 21, 2000.
24. Lauren Neergaard, "WHO issues antibiotic alert," Associated Press, AOL-online, June 12, 2000.
25. See Levy, *Antibiotic Paradox,* 113.
26. Quoted in Buhner, *Herbal Antibiotics,* 14.
27. Andreas Hartmann, et al. "Identification of fluoroquinone antibiotics as the main source of genotoxicity in native hospital wastewater," *Environmental Toxicology and Chemistry* 17, no. 3 (1998): 377–82; Peter Montague, "Drugs in the Water," *Rachel's #614*, September 3, 1998; Lappé, *When Antibiotics Fail*; Levy, *The Antibiotic Paradox;* Raloff, "More waters test positive for drugs."
28. Anita Manning, "Antibacterial soaps may create new problems," *USA Today,* September 22, 1998, 6D.
29. Maggie Fox, "Modern life comes with disease price tag," Reuters, AOL-online, September 20, 2000; Elsas and Smalla, "Antibiotic resistance traits in the environment."
30. Levy, *The Antibiotic Paradox*.
31. Ibid.
32. Quoted in Buhner, *Herbal Antibiotics,* 12–13.
33. Elsas and Smalla, "Antibiotic resistance traits in the environment."
34. Daughton and Ternes, "Pharmaceuticals and personal care products in the environment," 27; see also Janet Raloff, "Waterways carry antibiotic resistance," *Science News* 155 No. 23 (June 5, 1999).
35. Daughton and Ternes, ibid.
36. Levy, *The Antibiotic Paradox,* 175.
37. Elsas and Smalla, "Antibiotic resistance traits in the environment."
38. Ibid.; R.P. Bardos, "Coincidence of resistance to antibiotics and heavy metals in soil bacteria," *Journal of the Science of Food and Agriculture* 36 (1985): 539–40.
39. Pamela McCall, "Bitter pill for U.S. health," CBS News, June 5, 2000, www.cbs.aol.com.
40. Peter Montague, "Another kind of drug problem," *Rachel's* #632, January 7, 1999; Melody Peterson, "What's black and white and sells medicine?" *New York Times,* August 27, 2000, section 3, 1.
41. "New Drugs and the Media," MayoClinic, www.mayohealth.org, June 2, 2000.
42. John Abraham, *Science, Politics and the Pharmaceutical Industry: Controversy and Bias in Drug Regulation* (New York: St. Martins, 1995).
43. Montague, "Another kind of drug problem."
44. *USA Today,* April 24, 1998, 18A, editorial: "Deadly Rx: Why are drugs killing so many patients?"; Montague, ibid., *Journal of the American Medical Association* 279, no. 15 (April 15, 1998): 1200–1205.

45. Jon Cohen, "The hunt for the origin of AIDS," part 2, *Atlantic Monthly,* www.theatlantic.com, October 2000, 3.
46. Debbie Bookchin and Jim Schumacher, "The virus and the vaccine," *Atlantic Monthly,* February 2000, www.theatlantic.com.
47. Louis Pascal, "What Happens When Science Goes Bad," Science and Technology Analysis Research Programme, Working Paper No. 9 (Wollongong, Australia, University of Wollongong, December 1991), 13.
48. Tom Curtis, "The origin of AIDS," *Rolling Stone* 626 (March 19, 1992), 54.
49. Lynn Margulis, *Symbiotic Planet,* 75.
50. Schmidt, et al., *Beyond Antibiotics;* Lappé, *When Antibiotics Fail.*
51. Linda Johnson, "Germ exposure may be good for kids," Associated Press, AOL online, August 24, 2000.
52. "We need the worms," *Discover* magazine online, www.discover.com/search/index.html.
53. Lappé, *When Antibiotics Fail,* xviii.
54. Ibid., 25–26.
55. Ibid.

Chapter Seven

1. Jonathan Horton and Stephen Hart, "Hydraulic lift: A potentially important ecosystem process," *TREE* 13, no. 6 (June 1998): 232–35.
2. Alan Putnam and Chung-shih Tang, "Allelopathy: The state of the science," *The Science of Allelopathy,* ed. Alan Putnam and Chung-shih Tang (New York: John Wiley and Sons, 1986), 4.
3. David Hoffmann, *Phytochemistry: Molecular Veriditas,* work in progress, 31.
4. Frank Einhellig, "Mechanisms and modes of action of allelochemicals," and Putnam and Tang, "Allelopathy," both in *The Science of Allelopathy.*
5. Putnam and Tang, ibid.
6. John Lovett, "Allelopathy: The Australian experience," in *The Science of Allelopathy.*
7. May Berenbaum, "The chemistry of defense," in *Chemical Ecology,* ed. Thomas Eisner and Jerrold Meinwald (Washington, D.C.: National Academy of Sciences, 1995); see also J.P. Grime, et al., "Functional types: Testing the concept in northern England," in *Plant Functional Types; Their Relevance to Ecosystem Properties and Global Change,* ed. I.M. Smith, H.H. Shugart, and F.I. Woodward (Cambridge: Cambridge University Press, 1997).
8. Putnam and Tang, "Allelopathy," in *The Science of Allelopathy.*
9. Kenneth Stevens, "Polyacetylenes as allelochemicals," in ibid. 219–20.
10. Nicholaus Fischer and Leovigildo Quanijano, "Allelopathic agents from common weeds," in *The Chemistry of Allelopathy: Biochemical Interactions Among Plants,* ed. Alonzo Thompson (Washington, D.C.: American Chemical Society, 1985).
11. R.E. Hoagland and R.D. Williams, "The influence of secondary plant compounds on the associations of soil microorganisms and plant roots," in ibid., F.A. Einhellig, M. Stille Muth, and M.K. Schon, "Effects of allelochemicals on plant-water relationships," in ibid.; Paul Keddy and Lauchlan Fraser, "On the diversity of land plants," in *Ecoscience* 6, no. 3 (1999): 366–80.

12. Nelson Blake, "Effects of allelochemicals on mineral uptake and associated physiological processes," in *The Chemistry of Allelopathy*.
13. G.R. Waller and F.A. Einhellig, "Overview of allelopathy in agriculture, forestry, and ecology," in *Biodiversity and Allelopathy,* ed. Chang-Hung Chou, George Waller, and Charlie Reinhardt (Taipei, Taiwan: Academia Sinica, 1999).
14. Fischer and Quanijano, "Allelopathic agents from common weeds."
15. John Romeo, "Functional multiplicity among nonprotein amino acids in Mimosoid legumes: A case against redundancy," in *Ecoscience* 5, no. 3 (1998): 287–94.
16. Hoagland and Williams, "Soil microorganisms-plant root associations"; Lovett, "Allelopathy: The Australian experience."
17. Chung-Shih Tang, "Continuous trapping techniques for the study of allelochemicals from higher plants," in *The Science of Allelopathy*.
18. Hoagland and Williams, "Soil microorganisms-plant root associations."
19. J.B. Harborne, "Role of phenolic secondary metabolites in plants and their degradation in nature," in *Driven by Nature: Plant Litter Quality and Decomposition*, ed. G. Cadisch and K.E. Giller (Wallingford, England: CAB International, 1997).
20. A.R. Watkinson, "The role of the soil community in plant population dynamics," in *TREE* 13, no. 5 (May 1998): 171–72.
21. Ibid.; Lovett, "Allelopathy: The Australian experience."
22. Alan Putnam and Leslie Weston, "Adverse impact of allelopathy in agricultural systems," in *The Science of Allelopathy*.
23. Watkinson, "The role of the soil community in plant population dynamics." See, for example, witchweed germination triggers, William Agosta, *Bombardier Beetles and Fever Trees* (New York: Addison Wesley, 1996); 184, 195–96.
24. Watkinson, ibid.; S.K. Chile, "Efficacy of the extracts of some fleshy fungi against *Phytophthora parasitica* var. *piperina*," in *Herbal Medicines, Biodiversity and Conservation Strategies*, ed. R.C. Rajak and M.K. Rai (Dehra Dun, India: International Book Distributors, 1996), 61–69; Wade Davis, *Rainforest* (White River Junction, Vt.: Chelsea Green, 1998).
25. Davis, ibid; C. Reinhardt, et al., "Implications of allelopathic weeds for sustainability in agronomy, agroforestry, and forestry," in *Biodiversity and Allelopathy*.
26. Ibid.; Waller and Einhellig, "Overview of allelopathy in agriculture, forestry, and ecology."
27. Ibid.; Chang-Hung Chou, "Frontiers of allelopathy in sustainable agriculture: Experiences from Taiwan," in *Biodiversity and Allelopathy*.
28. E.A. Golovko, "Allelopathic soil sickness: Basic and methodological aspects," in *Biodiversity and Allelopathy.*
29. J.B. Harborne, *Introduction to Ecological Biochemistry* (London: Academic Press, 1982), see especially chapter 9, 227–59.
30. Hiroyuki Nishimura, et al., "Ecochemicals from chicory rhizome," in *Biodiversity and Allelopathy.*
31. Harborne, *Introduction to Ecological Biochemistry,* see 227–59.
32. Ibid, 235.

33. Ibid, 242; Shigeru Tamogami and Osamu Kodamal, "Jasmonic acid elicits momilactone A production: Physiology and chemistry of jasmonic acid in rice phytoalexin production," in *Biodiversity and Allelopathy*.
34. Einhellig, "Mechanisms and modes of action of allelochemicals."
35. Logical Nonsense, www.lynx.bc.ca/~jc/text.html, citing the July 1998 *Journal of Biological Chemistry* and the November-December 1998 issue of *The Sciences*; Tamogami and Kodamal, "Jasmonic acid elicits momilactone A production."
36. Einhellig, "Mechanisms and modes of action of allelochemicals."
37. Ibid.
38. Harborne, *Introduction to Ecological Biochemistry,* see especially chapter 3, 66–95.
39. Sharon Strauss and Anurag Agarwal, "The ecology and evolution of plant tolerance to herbivory," in *TREE* 14, no. 5 (May 1999): 179–85.
40. Ibid.; Tuija Honkanen and Erkki Haukioja, "Intra-plant regulation of growth and plant-herbivore interactions," in *Ecoscience* 5, no. 4 (1998): 470–79.
41. Strauss and Agarwal, "The ecology and evolution of plant tolerance to herbivory"; Harborne, *Introduction to Ecological Biochemistry,* 149.
42. Ibid, chapter 6, e.g., 161.
43. Harborne, "Role of phenolic secondary metabolites."
44. Harborne, *Introduction to Ecological Biochemistry,* 160, 137–41.
45. Romeo, "Functional multiplicity among nonprotein amino acids in Mimosoid legumes"; M. Deane Bowers, "Chemistry and coevolution: Iridoid glycosides, plants and herbivorous insects," in *Chemical Mediation of Coevolution,* ed. Kevin Spencer (San Diego: Academic Press, 1988).
46. Richard Karban, et al., "Induced plant responses and information content about risk of herbivory," in *TREE* 14, no. 11 (November 1999): 443–47.
47. Harborne, *Introduction to Ecological Biochemistry,* 83.
48. Tang, "Continuous trapping techniques for the study of allelochemicals from higher plants."
49. Harborne, *Introduction to Ecological Biochemistry,* 150.
50. Rex Cates and Richard Redak, "Variation in the terpene chemistry of Douglas-fir and its relationship to western spruce budworm success," in *Chemical Mediation of Coevolution.*
51. Ibid.
52. Romeo, "Functional multiplicity among nonprotein amino acids in Mimosoid legumes."
53. Harborne, *Introduction to Ecological Biochemistry,* especially chapter 4, 98–119.
54. Peter Montague, "Girls are reaching puberty early," *Rachel's* #566, October 2, 1997.
55. Plant Metabolism and Secondary Plant Compounds online, www.esb.utexas.edu/palmer/bio303/group14/background.html; Harborne, *Introduction to Ecological Biochemistry.*
56. Ibid.
57. Mark Stowe, "Chemical mimicry," in *Chemical Mediation of Coevolution.*
58. Mark Stowe, et al., "The chemistry of eavesdropping, alarm, and deceit," in

Chemical Ecology: The Chemistry of Biotic Interaction, ed. Thomas Eisner and Jerrold Meinwald (Washington, D.C.: National Academy of Sciences, 1995).

59. Agosta, *Bombardier Beetles and Fever Trees.*
60. Stowe, et al. "The chemistry of eavesdropping, alarm, and deceit."
61. Ted Turlings and Betty Benrey, "Effect of plant metabolites on the behavior and development of parasitic wasps," in *Ecoscience* 5, no.3 (1998): 321–33.
62. Brian Hocking, "Ant-plant mutualisms," in *Coevolution of Animals and Plants*, ed. Lawrence Gilbert and Peter Raven (Austin: University of Texas Press, 1980).
63. David Wilkinson, "Ants, agriculture, and antibiotics," in *TREE* 14, no.12 (December, 1999): 459–60.
64. O.W. Heal, et al., "Plant litter quality and decomposition: An historical overview," in *Driven by Nature.*
65. Benoit Cote and James Fyles, "Leaf Litter disappearance of hardwood species in southern Quebec: Interaction between litter quality and stand type," in *Ecoscience* 1, no. 4 (1994): 322–328.
66. Heal, et al., "Plant Litter Quality and Decomposition;" Joaquim Esteves da Silva, et al., "Method for the differentiation of leaf litter extracts and study of their interaction with CU(II) by molecular fluorescence," *Canadian Journal of Chemistry* 76, no. 8 (August 1998).
67. Chou, "Frontiers of Allelopathy in Sustainable Agriculture."
68. Chang-Hung Chou, "Allelopathy in Subtropical Agroecosystems in Taiwan," in *The Science of Allelopathy.*
69. Heal, et al., "Plant Litter Quality and Decomposition"; William Currie, "The responsive C and N biogeochemistry of the temperate forest floor," in *TREE* 14, no. 8 (August 1999).
70. "Leaf litter for healthy streams," *Forestry Notes,* www.nacdnet.org/forestrynotes/april98/streams.htm.
71. Eun Ju Lee, et al., "Atmospheric deposition of macronutrients by pollen in the boreal forest," in *Ecoscience* 3, no. 3 (1996): 304–306; Elroy Rice, "Allelopathic growth stimulation," in *The Science of Allelopathy.*
72. Rice, ibid.
73. See, for example, Alejo Carballeira and Manuel Reigosa, "Allelopathic natural leachates of *Acacia dealbata,*" in *Bot Bull Acad Sin* 40 (1999): 87–92, http://ejournal.sinica.edu.tw.
74. D.A. Wardle and P. Lavelle, "Linkages between soil biota, plant litter quality, and decomposition," in *Driven by Nature.*
75. Johannes Knops, et al., "The influence of epiphytic lichens on the nutrient cycling of an oak wood," in *Ecological Monographs* (May 1996) no vol. or issue given, www.northernlight.com.
76. T.E. Dawson, "Fog in the California redwood forest: Ecosystem inputs and use by plants," in *Oecologia* 117 (spring 1998): 476–85; Kathleen Weathers, "The importance of cloud and fog in the maintenance of ecosystems," in *TREE* 14, no. 6 (1999).
77. Randal Buddington, "The use of fermentable fibers to manage the gastrointestinal ecosystem," in *Phytochemicals as Bioactive Agents,* ed. Wayne Bidlack, et al. (Lancaster, Pa.: Technomic Publishing, 2000); Richard

Lindroth, "Adaptations of mammalian herbivores to plant chemical defenses," in *Chemical Mediation of Coevolution;* Wardle and Lavelle, "Linkages between soil biota, plant litter quality, and decomposition."

78. Lovett, "Allelopathy: The Australian experience."
79. Patricia Lauber, *Earthworms* (New York: Henry Holt, 1994); Charles Darwin, *The Formation of Vegetable Mould Through the Action of Worms With Observations on Their Habits* (reprint, London: Bookworm Publishing, 1976); Mary Appelhof, ed., *Workshop on the Stabilization of Organic Residues,* volume 1 (Kalamazoo, Mich.: Beech Leaf Press, 1981); G. Tian, et al., "Soil fauna-mediated decomposition of plant residues under constrained environmental and residue quality conditions," in *Driven by Nature.*
80. C. Wachendorf, et al., "Relationships between litter fauna and chemical changes of litter during decomposition under different moisture conditions," in *Driven by Nature.*
81. D. Wardle and P. Lavelle, "Linkages between soil biota, plant litter quality, and decomposition," in *Driven by Nature.*
82. James Lovelock, *Healing Gaia* (New York: Harmony Books, 1991).
83. Peter Tompkins and Christopher Bird, *The Secret Life of Plants* (New York: Harper and Row, 1973).
84. Einhellig, "Mechanisms and modes of action of allelochemicals."
85. Nikolaus Fisher, "Mono and sesquiterpenes as plant regulators," in *The Science of Allelopathy.*

Chapter Eight

1. James Lovelock, *Healing Gaia,* 25.
2. K.D. Bennett, "The power of movement in plants," in *TREE* 13, no. 9 (September 1998): 339.
3. See, for example, Alberto Burquez and Maria de los Angeles Quintana, "Diversity under ironwood," in *Ironwood: An Ecological and Cultural Keystone of the Sonoran Desert,* ed. Gary Paul Nabhan and John L. Carr (Washington, D.C.: Conservation International, Occasional Paper No. 1, April 1994); Bennett, "The power of movement in plants."
4. H.H. Shugart, "Plant and ecosystem functional types," in *Plant Functional Types: Their Relevance to Ecosystem Properties and Global Change,* T.M. Smith, H.H Shugart, and F.I. Woodward, eds. (Cambridge: Cambridge University Press, 1997).
5. Gregory Bateson, *Mind and Nature,* 16–17.
6. O.W. Heal and J.P. Grime, 1991, quoted in H. Gitay and I.R. Noble, "What are functional types and how should we seek them?" in *Plant Functional Types,* 3.
7. R.J. Scholes, et al., "Plant functional types in African savannas and grasslands," in *Plant Functional Types,* 255.
8. Burquez and Quintana, "Diversity under ironwood."
9. Nabhan and Carr, *Ironwood.*
10. Frank Einhellig, "Mechanisms and modes of action of allelochemicals," in *The Science of Allelopathy.*
11. Burquez and Quintana, "Diversity under ironwood," 22.
12. Shahid Naeem, et al., "Plant neighborhood diversity and production," in

Ecoscience 3, no. 3 (1999): 355–65; J.P. Grime, "Benefits of plant diversity to ecosystems: Intermediate, filter, and founder effects," in *Journal of Ecology* 86 (1998): 902–10.
13. Steve Archer, "Tree-grass dynamics in a *Prosopis*-thornscrub savanna parkland: Reconstructing the past and predicting the future," in *Ecoscience* 1, no. 1 (1995): 83–99.
14. Grime, "Benefits of plant diversity to ecosystems."
15. Bennett, "The power of movement in plants," 339.
16. Ragan Callaway, "Positive interactions among plants," in *The Botanical Review* 61, no. 4 (October 12, 1995).
17. Naeem, et al., "Plant neighborhood diversity and production," 355–365; Grime, "Benefits of plant diversity to ecosystems."
18. Grime, ibid., 907.
19. Naeem, et al., ibid.; Grime, ibid.
20. Ibid.; M.W. Rode, "Leaf nutrient accumulation and turnover via litter fall in three stages of succession from heathland to forest," in *Journal of Vegetation Science* 4 (1993): 263–68.
21. Naeem, et al., "Plant neighborhood diversity and production." Callaway, "Positive interactions among plants."
22. Ibid.
23. Naeem, et al., "Plant neighborhood diversity and production."
24. Keith Clay, "Parallel worlds converge," in *TREE* 13, no. 1 (January 1998): 38–39; Callaway, "Positive interactions among plants."
25. Todd Dawson, "Agriculture in nature's image," in *TREE* 13, no. 2 (February 1998): 4–5; M. An, et al., "Allelopathy from concept to reality," http://life.csu.edu.au; Phillip Fujiyoshi, *Mechanisms of Weed Suppression by Squash Intercropped in Corn*, dissertation excerpt, http://agroecology.org/phillip.html; Fukuoka, *Natural Way of Farming*.
26. Buhner, *Sacred and Herbal Healing Beers*.
27. Fukuoka, *Natural Way of Farming*.
28. Gary Paul Nabhan and Mark Plotkin, "Introduction," in *Ironwood*.
29. Richard Gomulkiewicz, "A rare treat," in *TREE* 13, no. 1 (January 1998): 38.
30. Neil West and John Skujins, *Nitrogen in Desert Ecosystems* (Stroudsburg, Pa.: Dowden, Hutchinson and Ross, 1978).
31. W.K. Laurenroth, et al., "Interactions between demographic and ecosystem processes in a semi-arid grassland: A challenge for plant functional types," in *Plant Functional Types*; W.J. Bond, *Plant Functional Types,* 174.
32. Evelyn Fox Keller, *A Feeling for the Organism* (New York: W.H. Freeman, 1983), xix.
33. Richard Karban, et al., "Induced plant responses and information content about risk of herbivory," in *TREE* 14, no. 11 (1999); Lovelock, *Healing Gaia*.
34. Jonathan Silvertown, "Plant phenotype plasticity and non-cognitive behaviour," in *TREE* 13, no. 7 (July 1998): 255–56.
35. Lawrence Gilbert, "Ecological consequences of a coevolved mutualism between butterflies and plants," in Gilbert and Raven, eds., *Coevolution of Animals and Plants*.
36. Kevin Spencer, "Chemistry and coevolution," in Spencer, ed., *Chemical Mediation of Coevolution*; M. Deane Bowers, "Chemistry and coevolution:

Iridoid glycosides, plants and herbivorous insects," ibid.

37. Rex Cates and Richard Redak, "Variation in the terpene chemistry of Douglas-fir and its relationship to western spruce budworm success," in *Chemical Mediation of Coevolution;* Alan Gemmill and Andrew Read, "Counting the cost of disease resistance," in *TREE* 13, no. 1 (January 1998).
38. Harborne, *Introduction to Ecological Biochemistry*, 109–12.
39. Ibid.
40. William Agosta, *Bombardier Beetles and Fever Trees,* 166–67.
41. Gary Paul Nabhan, *The Forgotten Pollinators* (Washington, D.C.: Island Press, 1996), 61.
42. Tompkins and Bird, *The Secret Life of Plants,* 106.
43. Harborne, *Introduction to Ecological Biochemistry,* chapter 2, 32–64; Herbert Baker and Irene Baker, "Nectar constituents," in *Coevolution of Animals and Plants.*
44. Nabhan, *Forgotten Pollinators,* 92; see also Harborne, ibid.
45. Elisa Bernklau, "Chemical mimicry in pollination," www.colostate.edu/Depts/Entomology/courses/en570/papers_1996/bernklau.html; Nabhan, *Forgotten Pollinators,* 49–52; Mark Stowe, "Chemical mimicry," in *Chemical Mediation of Coevolution.*
46. Ibid.; Harborne, *Introduction to Ecological Biochemistry,* 57.
47. Ibid; Stowe, "Chemical mimicry."
48. Lawrence Gilbert, "Ecological consequences of a coevolved mutualism between butterflies and plants," in *Coevolution of Animals and Plants;* Kevin Spencer, "Chemical mediation of coevolution in the *Passiflora-Heliconium* interaction," in *Chemical Mediation of Coevolution.*
49. Nabhan, *Forgotten Pollinators,* 54–56.
50. O.W. Heal, et al., "Plant litter quality and decomposition: An historical overview," in *Driven by Nature*; David Perry, "A moveable feast: The evolution of resource sharing in plant-fungus communities," in *TREE* 13, no. 11 (November 1998).
51. Callaway, "Positive interactions among plants."
52. Harborne, *Introduction to Biological Chemistry,* 203; Clive Jones, et al., "Diet breadth and insect chemical defenses: A generalist grasshopper and general hypothesis," in *Chemical Mediation of Coevolution*.
53. Ibid, chapter 6, especially 193–204.
54. Agosta, *Bombardier Beetles and Fever Trees,* 42.
55. N.R. Salgado, et al., "Zoopharmacognosy of breeding neotropical birds: Chemical ecology of nest materials and substrates," Zoopharmacognosy abstracts, www.bhort.bh.cornell.edu; Judith Sumner, *The Natural History of Medicinal Plants* (Portland, Oreg.: Timber Press, 2000).
56. Agosta, *Bombardier Beetles and Fever Trees,* 99.
57. Online: http://journeytoforever.org/quackgrass.html.
58. Sumner, *The Natural History of Medicinal Plants.*
59. Valentina Carrai, "Increase in tannin uptake by *Propithecus verraeauxi verreauxi* (Sifaka) females during birth season: A possible case of zoopharmacognosy?" Notes from the Kirindy Symposium: www.dpz.gwdg.de/voe_page/kirindy-symposium.htm.
60. Eloy Rodriguez and Richard Wrangham, "Zoopharmacognosy: The use of

medicinal plants by animals," in *Phytochemical Potential of Tropical Plants*, ed. K.R. Downum, et al. (New York: Plenum Press, 1993); Michael Huffman, "The medicinal use of plants by chimpanzees in the wild," http://jinrui.zool.kyoto-u.ac.jp/CHIMPP/; Sumner, *The Natural History of Medicinal Plants*; Michael Huffman, et al. "Leaf-swallowing by chimpanzees: A behavioral adaptation for the control of strongyle nematode infections," in *International Journal of Primatology* 17, no. 4, 1996.

61. Shawn Sigstedt, *Bear Medicine: The Medical Use of Osha by Bears,* American Herbalist Guild Annual Conference, 1993 (audio tape, Tree Farm tapes, 206-868-0464); Rebecca Andrews, "Western science learns from native culture," in *The Scientist* 6, no. 6 (March 16, 1992).
62. Logical Nonsense: www.lynx.bc.ca/~jc/text.html, citing the July 1998 *Journal of Biological Chemistry* and the November-December 1998 issue of *The Sciences*.
63. Frietson Galis, "Ecology through the chemical looking-glass," in *TREE* 13, no. 1 (January 1998): 2–3.
64. Daughton and Ternes, "Pharmaceuticals and personal care products in the environment."
65. Agosta, *Bombardier Beetles and Fever Trees,* 90; Thomas Newmark and Paul Schilick, *Beyond Aspirin* (Prescott, Ariz.: Hohm Press, 2000).
66. Harborne, *Introduction to Ecological Biochemistry,* 220.
67. Daughton and Ternes, "Pharmaceuticals and personal care products in the environment."
68. Einhellig, "Mechanisms and modes of action of allelochemicals,"; G.R. Waller and F.A. Einhellig, "Overview of allelopathy in agriculture, forestry, and ecology," in *Biodiversity and Allelopathy.*
69. Harborne, *Introduction to Ecological Biochemistry,* 94.
70. A. Chesson, "Plant degradation by ruminants: Parallels with litter decomposition in soils," in *Driven by Nature.*
71. Tompkins and Bird, *Secret Life of Plants,* 225.
72. Timothy Johns and Laurie Chapman, "Phytochemicals ingested in traditional diets and medicines as modulators of energy metabolism," in *Phytochemistry of Medicinal Plants,* ed. John Amason, et al. (New York: Plenum Press, 1995).
73. Stephen Harrod Buhner, *Herbal Antibiotics*.
74. Robert McCaleb, "Medicinal plants for healing the planet: Biodiversity and environmental health care," in *Biodiversity and Human Health,* ed. Francesca Grifo and Joshua Rosenthal (Washington, D.C.: Island Press, 1997).
75. H.A. Mooney, "Ecosystem function of biodiversity," in *Plant Functional Types*.
76. Howard Odum, ed., *Heavy Metals in the Environment: Using Wetlands for Their Removal* (Boca Raton: Lewis Publishers, 2000); Elly Best, "Phytoremediation of explosives in groundwater using constructed wetlands," www.wes.army.mil.
77. Anna Joabsson, et al., "Vascular plant controls on methane emissions from northern peatforming wetlands," in *TREE* 14, no. 10 (October 1999).
78. Quoted in Timothy Coffee, *Wildflowers* (New York: Houghton Mifflin, 1993) 266.

Chapter Ten

1. Delores LaChappelle, *Sacred Land, Sacred Sex, Rapture of the Deep*, (Silverton, Colo.: Finn Hill Arts, 1988), 272.
2. Derrick Jensen, "The plants respond: An interview with Cleve Backster," *The Sun,* July 1997.
3. M. Deane Bowers, "Chemistry and coevolution: Iridoid glycosides, plants and herbivorous insects," in Spencer, ed., *Chemical Mediation of Coevolution.*
4. Wendell Berry, quoted in *HerbalGram* 26 (1992): 50.
5. See, for example, Hillman, *Soul's Code*, Ehrenfeld, *Arrogance of Humanism;* Bill Devall and George Sessions, *Deep Ecology* (Salt Lake City, Utah: Peregrine Smith Books, 1985), 182–83; Aldo Leopold, "Natural history," in *A Sand County Almanac*.
6. Robert Bly, *The Kabir Book* (Boston: Beacon Press, 1977), 7.

Bibliography

Abraham, John. *Science, Politics and the Pharmaceutical Industry: Controversy and Bias in Drug Regulation.* New York: St. Martins, 1995.

Abram, David. "Trust Your Senses," in *Only Connect: Soil, Soul, Society*, ed. John Lane and Maya Kumar Mitchell. White River Junction, Vt.: Chelsea Green, 2000.

Agosta, William. *Bombardier Beetles and Fever Trees*. New York: Addison Wesley, 1996.

Akerle, Olayiwola,. V. Heywood, and H. Synge, eds. *Conservation of Medicinal Plants.* Cambridge: Cambridge University Press, 1991.

Alibek, Ken. *Biohazard: The Chilling True Story of the Largest Covert Biological Weapons Program in the World Told From Inside By the Man Who Ran It.* New York: Random House, 1999.

Appelhof, Mary, ed. *Workshop on the Stabilization of Organic Residues,* volume 1. Kalamazoo, Mich.: Beech Leaf Press, 1981.

Archer, Steve. "Tree-grass dynamics in a *Prosopis*-thornscrub savanna parkland: Reconstructing the past and predicting the future," *Ecoscience* 1, no. 1, 1995.

Arnason, John, Rachel Mata, and John Romeo, eds. *Phytochemistry of Medicinal Plants*. New York: Plenum Press, 1995.

Ausubel, Kenny. *When Healing Becomes a Crime*. Rochester, Vt.: Healing Arts Press, 2000.

Baker, Herbert, and Irene Baker. "Nectar constituents," in *Coevolution of Animals and Plants*, ed. Lawrence Gilbert and Peter Raven. Austin: University of Texas Press, 1980.

Bateson, Gregory. *Mind and Nature: A Necessary Unity*. New York: E.P. Dutton, 1979.

Bennett, K.D. "The power of movement in plants," *TREE* 13, no. 9, September 1998.

Berenbaum, May. "The chemistry of defense," in Mark Stowe, et al., eds., *Chemical Ecology.* Washington D.C.: National Academy of Sciences, 1995.

Berry, Wendell. *The Unsettling of America*. San Francisco: Sierra Club Books, 1977.

Beston, Henry. *Herbs and the Earth*. Boston: David Godine, 1990.

Bidlack, Wayne, S. Omaye, M. Meskin, and D. Topham, eds. *Phytochemicals as Bioactive Agents.* Lancaster, Pa.: Technomic Publishing, 2000.

Blake, Nelson. "Effects of allelochemicals on mineral uptake and associated physiological processes," in *The Chemistry of Allelopathy: Biochemical*

Interactions Among Plants, ed. Alonzo Thompson. Washington, D.C.: American Chemical Society, 1985.

Bly, Robert. *The Winged Life*. San Francisco: Sierra Club Books, 1986.

Bond, W.J. "Functional types for predicting changes in biodiversity: A case study in Cape fynbos," in *Plant Functional Types: Their Relevance to Ecosystem Properties and Global Change*, ed. T.M. Smith, H.H Shugart, and F.I. Woodward. Cambridge: Cambridge University Press, 1997.

Bookchin, Debbie, and Jim Schumacher. "The virus and the vaccine," *The Atlantic Monthly Online*, www.theatlantic.com, February 2000.

Bowers, M. Deane. "Chemistry and coevolution: Iridoid glycosides, plants and herbivorous insects," in *Chemical Mediation of Coevolution*, ed. Kevin Spencer. New York: Academic Press, 1988.

Buddington, Randal. "The use of fermentable fibers to manage the gastrointestinal ecosystem," in *Phytochemicals as Bioactive Agents*, ed. Wayne Bidlack, et al. Lancaster, Pa.: Technomic Publishing, 2000.

Buhner, Stephen Harrod. *Herbal Antibiotics: Natural Alternatives for Drug-resistant Bacteria*. Pownal, Vt.: Storey, 1999.

———. *Sacred and Herbal Healing Beers: The Secrets of Ancient Fermentation*. Boulder, Colo.: Siris Books, 1998.

———. *One Spirit, Many Peoples*. Boulder, Colo.: Roberts Rinehart, 1997.

———. *Sacred Plant Medicine*. Boulder,Colo.: Roberts Rinehart, 1996.

Cadisch, G., and K.E. Giller, eds. *Driven by Nature: Plant Litter Quality and Decomposition*. Wallingford, England: CAB International, 1997.

Callaway, Ragan. "Positive interactions among plants," *The Botanical Review* 61, no. 4, October 12, 1995.

Cates, Rex, and Richard Redak. "Variation in the terpene chemistry of Douglas-fir and its relationship to western spruce budworm success," in *Chemical Mediation of Coevolution*, ed. Kevin Spencer. New York: Academic Press, 1988.

Chase, Alston. "Harvard and the making of the Unibomber," *The Atlantic Monthly*, June 2000.

Chesson, A. "Plant degradation by ruminants: parallels with litter decomposition in soils," in *Driven by Nature*, ed. G. Cadisch and K.E. Giller. Wallingford, England: CAB International, 1997.

Chile, S.K. "Efficacy of the extracts of some fleshy fungi against *Phytophthora parasitica* var. *piperina*," in *Herbal Medicines, Bio-Diversity and Conservation Strategies*, ed. R.C. Rajak and M.K. Rai. Dehra Dun, India: International Book Distributors, 1996.

Chou, Chang-Hung. "Frontiers of allelopathy in sustainable agriculture: Experiences from Taiwan," in *Biodiversity and Allelopathy*, ed. Chang-Hung Chou, George Waller, and Charlie Reinhardt. Taipei, Taiwan: Academia Sinica, 1999.

———. "Allelopathy in subtropical agroecosystems in Taiwan," in *The Science of Allelopathy*, ed. Alan Putnam and Chung-shih Tang. New York: John Wiley and Sons, 1986.

Clay, Keith. "Parallel worlds converge," *TREE* 13, no. 1, January 1998.

Cohen, Jon. "The hunt for the origin of AIDS," *The Atlantic Monthly Online*, www.theatlantic.com, October 2000, in 3 parts.

Cote, Benoit, and James Fyles. "Leaf litter disappearance of hardwood species in southern Quebec: Interaction between litter quality and stand type," *Ecoscience* 1, no. 4, 1994.
Curtis, Tom. "The origin of AIDS: A startling new theory attempts to answer the question 'was it an act of God or an act of man?'" *Rolling Stone* 626, March 19, 1992. Online at: www.uow.edu.au/arts/sts/bmartin/dissent/documents.AIDS/Pascal91.html
Darwin, Charles. *The Formation of Vegetable Mould Through the Action of Worms With Observations on Their Habits.* London: Bookworm Publishing, 1976.
Daughton, Christian, and Thomas Ternes. "Pharmaceuticals and personal care products in the environment: Agents of subtle change?," *Environmental Health Perspectives* 107, Supplement 6, December 1999.
Davis, Wade. *Rainforest*. White River Junction, Vt.:Chelsea Green, 1998.
Dawson, Todd. "Agriculture in nature's image," *TREE* 13, no. 2, February 1998.
Dawson, T.E. "Fog in the California redwood forest: Ecosystem inputs and use by plants," in *Oecologia* 117, Spring 1998.
DeMaille, Raymond, ed. *The Sixth Grandfather: Black Elk's Teachings Given to John Neihardt*. Lincoln: University of Nebraska Press, 1985.
Densmore, Francis. *Teton Sioux Music*. Washington, D.C.: Smithsonian Institution, Bureau of American Ethnology, Bulletin 61, 1918.
Dicke, Marcel. "Plants in action," *TREE* 13, no. 2, February 1998.
Dickinson, C.H., and G.J.F. Pugh. *Biology of Plant Litter Decomposition*. 2 vols. New York: Academic Press, 1974.
Downum, Kelsey R., John Romeo, and Helen Stafford. *Phytochemical Potential of Tropical Plants.* New York: Plenum Press, 1993.
Duke, James. *Handbook of Phytochemical Constituents of GRAS Herbs and Other Economic Plants*. Boca Raton, Fla.: CRC Press, 1992.
———. "*Herbal Voices* interview with Jim Duke," *Herbal Voices* 2, no. 2, 2000.
Ehrenfeld, David. *The Arrogance of Humanism*. New York: Oxford University Press, 1978.
Einhellig, Frank. "Mechanisms and modes of action of allelochemicals," in *The Science of Allelopathy,* ed. Alan Putnam and Chung-shih Tang. New York: John Wiley and Sons, 1986.
———, M. Stille Muth, and M.K. Schon. "Effects of allelochemicals on plant-water relationships," in *The Chemistry of Allelopathy: Biochemical Interactions Among Plants*, ed. Alonzo Thompson. Washington, D.C.: American Chemical Society, 1985.
Eliade, Mircea. *The Forbidden Forest*. Notre Dame, Ind.: University of Notre Dame, 1978.
Fagin, Dan, Marianne Lavelle, and the Center for Public Integrity. *Toxic Deception: How the Chemical Industry Manipulates Science, Bends the Law and Endangers Your Health*. Monroe, Maine: Common Courage Press, 1999.
Fisher, Nikolaus. "Mono and sesquiterpenes as plant regulators," in *The Science of Allelopathy,* ed. Alan Putnam and Chung-shih Tang, New York: John Wiley and Sons, 1986.
———, and Leovigildo Quanijano, "Allelopathic agents from common weeds," in *The Chemistry of Allelopathy: Biochemical Interactions Among Plants*, ed.

Alonzo Thompson. Washington, D.C.: American Chemical Society, 1985.
Fukuoka, Masanobu. *The Road Back to Nature*. New York: Japan Publications, 1987.
———. *The Natural Way of Farming: The Theory and Practice of Green Philosophy*. New York: Japan Publications, 1985.
Galis, Frietson. "Ecology through the chemical looking-glass," *TREE* 13, no. 1, January 1998.
Gardner, John. *The Art of Fiction: Notes on Craft for Young Writers*. New York: Vintage Books, 1991.
Gemmill, Alan, and Andrew Read. "Counting the cost of disease resistance," *TREE* 13, no. 1, January 1998.
Gilbert, Lawrence. "Ecological consequences of a coevolved mutualism between butterflies and plants," in *Coevolution of Animals and Plants,* ed. Lawrence Gilbert and Peter Raven, Austin: University of Texas Press, 1980.
Gitay, H., and I.R. Noble. "What are functional types and how should we seek them?," in *Plant Functional Types: Their Relevance to Ecosystem Properties and Global Change,* ed. T.M. Smith, H.H. Shugart, and F.I. Woodward. London: Cambridge University Press, 1997.
Gofman, John. *Radiation From Medical Procedures in the Pathogenesis of Cancer and Ischemic Heart Disease*. San Francisco: Committee for Nuclear Responsibility, 1999.
Golovko, E.A. "Allelopathic soil sickness: Basic and methodological aspects," in *Biodiversity and Allelopathy,* ed. Chang-Hung Chou, George Waller, and Charlie Reinhardt. Taipei, Taiwan: Academia Sinica, 1999.
Grifo, Francesca, and Joshua Rosenthal, eds. *Biodiversity and Human Health*. Washington D.C.: Island Press, 1997.
Griggs, Barbara. *Green Pharmacy*. Rochester, Vt.: Healing Arts Press, 1991.
Grime, J.P. "Benefits of plant diversity to ecosystems: Intermediate, filter, and founder effects," *Journal of Ecology* 86, 1998.
———, et al. "Functional types: Testing the concept in northern England," in *Plant Functional Types: Their Relevance to Ecosystem Properties and Global Change,* ed. T. M. Smith, H. H. Shugart, and F. I. Woodward. London: Cambridge University Press, 1997.
Grossinger, Richard. *Planet Medicine*. Boulder, Colo.: Shambhala, 1982.
Harborne, J.B. "Role of phenolic secondary metabolites in plants and their degradation in nature," in *Driven by Nature: Plant Litter Quality and Decomposition*, ed. G. Cadisch and K.E. Giller. Wallingford, England: CAB International, 1997.
———. *Introduction to Ecological Biochemistry*. London: Academic Press, 1982.
Heal, O.W., et al. "Plant litter quality and decomposition: An historical overview," in *Driven by Nature: Plant Litter Quality and Decomposition*, ed. G. Cadisch and K.E. Giller. Wallingford, England: CAB International, 1997.
Health Care Without Harm. *Greening Hospitals: An Analysis of Pollution Prevention in America's Top Hospitals*. Falls Church, Va.: Center for Environment and Justice, 1998.
———. *Mercury Thermometers and Your Family's Health*. Center for Health, Environment and Justice: Falls Church, Va., nd.

Hillman, James. *The Soul's Code*. New York: Random House, 1996.
Hoagland, R.E., and R.D. Williams. "The influence of secondary plant compounds on the association of soil microorganisms and plant roots," in *The Chemistry of Allelopathy: Biochemical Interactions Among Plants,* ed. Alonzo Thompson. Washington, D.C.: American Chemical Society, 1985.
Hocking, Brian. "Ant-plant mutualisms," in *Coevolution of Animals and Plants,* ed. Lawrence Gilbert and Peter Raven. Austin: University of Texas Press, 1980.
Hoffmann, David. *Phytochemistry: Molecular Veriditas.* In progress, n.p., n.d.
Honkanen, Tuija, and Erkki Haukioja. "Intra-plant regulation of growth and plant-herbivore interactions," *Ecoscience* 5, no. 4, 1998.
Horton, Jonathan, and Stephen Hart. "Hydraulic lift: A potentially important ecosystem process," *TREE* 13, No. 6, June 1998.
Huffman, Michael. "The medicinal use of plants by chimpanzees in the wild," online at http://jinrui.zool.kyoto-u.ac.jp/CHIMPP.
——— et al. "Leaf-swallowing by chimpanzees: A behavioral adaptation for the control of strongyle nematode infections," *International Journal of Primatology* 17, no. 4, 1996.
Jablonka, Eva, Marion Lamb, and Eytan Avital. "'Lamarckian' mechanisms in Darwinian evolution," *TREE* 13, no. 5, May 1998.
Jensen, Derrick. "The Plants Respond: An Interview with Cleve Backster," *The Sun,* July 1997.
Johns, Timothy, and Laurie Chapman. "Phytochemicals ingested in traditional diets and medicines as modulators of energy metabolism," in *Phytochemistry of Medicinal Plants,* ed. John Arnason, Rachel Mata, and John Romeo. New York: Plenum Press, 1995.
Jones, Clive, et al. "Diet breadth and insect chemical defenses: A generalist grasshopper and general hypothesis," in *Chemical Mediation of Coevolution,* ed. Kevin Spencer. New York: Academic Press, 1988.
Karban, Richard, et al. "Induced plant responses and information content about risk of herbivory," *TREE* 14, no. 11, November 1999.
Keddy, Paul, and Lauchlan Fraser. "On the diversity of land plants," *Ecoscience* 6, no. 3, 1999.
Kellert, Stephen. "The biological basis for human values of nature," in *The Biophilia Hypothesis,* ed. Edward O. Wilson and Stephen Kellert. Washington, D.C.: Island Press, 1993.
Knops, Johannes, et al. "The influence of epiphytic lichens on the nutrient cycling of an oak wood," *Ecological Monographs* (online: northernlight.com), May 1996.
Kolata, Gina. "A question of beauty: Is it good for you?" *New York Times,* June 13, 1999.
LaChappelle, Delores. *Sacred Land, Sacred Sex, Rapture of the Deep.* Silverton, Colo.: Finn Hill Arts, 1988.
Lamb, R. Bruce. *Rio Tigre and Beyond.* Berkeley, Calif.: North Atlantic Books, 1985.
———. *Wizard of the Upper Amazon.* Boston: Houghton Mifflin, 1974.
Lappé, Marc. *When Antibiotics Fail.* Berkeley, Calif.: North Atlantic Press, 1986.

Lauber, Patricia. *Earthworms*. New York: Henry Holt, 1994.
Laurenroth, W.K., et al., "Interactions between demographic and ecosystem processes in a semi-arid grassland: A challenge for plant functional types," in *Plant Functional Types: Their Relevance to Ecosystem Properties and Global Change,* ed. T.M. Smith, H.H. Shugart, and F.I. Woodward. London: Cambridge University Press, 1997.
Lee, Eun Ju, et al. "Atmospheric deposition of macronutrients by pollen in the boreal forest," *Ecoscience* 3, no. 3, 1996.
Leigh, Egbert Giles. "The modern synthesis, Ronald Fisher and creationism," *TREE* 14, no. 12, December 1999.
Lenssen, Nicholas. *Nuclear Waste: The Problem That Won't Go Away*. Washington, D.C.: Worldwatch Paper 106, December 1991.
Leopold, Aldo. *A Sand County Almanac*. 1949. reprint, New York: Ballantine, 1991.
Levy, Stuart. *The Antibiotic Paradox*. New York: Plenum Press, 1992.
Lindroth, Richard. "Adaptations of mammalian herbivores to plant chemical defenses," in *Chemical Mediation of Coevolution*, ed. Kevin Spencer. New York: Academic Press, 1988.
Lopez, Barry. "The Language of Animals," in *Only Connect: Soil, Soul, Society*, ed. John Lane and Maya Kumar Mitchell. White River Junction, Vt.: Chelsea Green, 2000.
Lovelock, James. *Healing Gaia: Practical Medicine for the Planet*. New York: Harmony Books, 1991.
Lovett, John. "Allelopathy: The Australian experience," in *The Science of Allelopathy,* ed. Alan Putnam and Chung-shih Tang. New York: John Wiley and Sons, 1986.
Margulis, Lynn. *Symbiotic Planet: A New Look at Evolution*. New York: Basic Books, 1998.
———, Karlene Schwartz, and Michael Dolan. *Diversity of Life: The Illustrated Guide to the Five Kingdoms*. Sudbury, Mass.: Jones and Bartlett, 1999.
———, and Dorion Sagan, *What Is Life?* New York: Simon and Schuster, 1995.
———, and Lorraine Olendzenski. *Environmental Evolution: Effects of the Origin and Evolution of Life on Planet Earth.* Cambridge: MIT Press, 1992.
———, and René Fester. *Symbiosis as a Source of Evolutionary Innovation: Speciation and Morphogenesis.* Cambridge: MIT Press, 1991
Martin, Brian. "Scientific proof and the origin of AIDS," online at www.uow.edu.au/arts/sts/bmartin/dissent/documents/AIDS.
Milton, Richard. *Shattering the Myths of Darwinism*. Rochester, Vt.: Park Street Press, 1997.
Mollison, Bill. *Permaculture: A Practical Guide for a Sustainable Future*. Washington, D.C.: Island Press, 1990.
Montague, Peter. Various articles in *Rachel's Health and Environment Weekly,* online at www.rachel.org.
Mooney, H.A. "Ecosystem function of biodiversity," in *Plant Functional Types: Their Relevance to Ecosystem Properties and Global Change,* ed. T.M. Smith, H.H. Shugart, and F.I. Woodward. London: Cambridge University Press, 1997.

Naeem, Shahid, et al. "Plant neighborhood diversity and production," *Ecoscience* 3, no. 3, 1999.
Nabhan, Gary Paul. *Cultures of Habitat*. Washington, D.C.: Counterpoint Press, 1997.
———, and John L. Carr, eds. *Ironwood: An Ecological and Cultural Keystone of the Sonoran Desert*. Washington D.C.: Conservation International, Occasional Paper No. 1, April 1994.
———. *The Forgotten Pollinators*. Washington, D.C.: Island Press, 1996.
———, and Sara St. Antoine. "The loss of floral and faunal story: The extinction of experience," in *The Biophilia Hypothesis,* ed. Edward O. Wilson and Stephen Kellert. Washington D.C.: Island Press, 1993.
Nelson, Richard. "Searching for the lost arrow: Physical and spiritual ecology in the hunter's world," in *The Biophilia Hypothesis,* ed. Edward O. Wilson and Stephen Kellert. Washington, D.C.: Island Press, 1993.
Nerburn, Kent, and Louise Mengelkoch, eds. *Native American Wisdom*. San Rafael, Calif.: New World Library, 1991.
Nishimura, Hiroyuki, et al. "Ecochemicals from chicory rhizome," in *Biodiversity and Allelopathy,* ed. Chang-Hung Chou, George Waller, and Charlie Reinhardt. Taipei, Taiwan: Academia Sinica, 1999.
Noss, Reed. "The naturalists are dying off," *Conservation Biology* 10, no. 1, February 1996.
Oliver, Mary. *House of Light*. Boston: Beacon Press, 1990.
Orr, David. "2020: A proposal," *Conservation Biology* 14, no. 2, April 2000.
———. "Verbicide," *Conservation Biology* 4, no.13, August 1999: 697.
———. "Virtual nature," *Conservation Biology* 10, no. 1, February, 1996.
———. *Earth in Mind: On Education, Environment, and the Human Prospect*. Washington, D.C.: Island Press, 1994.
———, and David Ehrenfeld. "None so blind: The problem of ecological denial," *Conservation Biology* 9, no. 5, October 1995.
Pace, Michael, J. Cole, S. Carpenter, and J. Kitchell. "Trophic cascades revealed in diverse ecosystems," *TREE* 14, no. 12, December 1999.
Pascal, Louis. "What happens when science goes bad: The corruption of science and the origin of AIDS: A study in spontaneous generation," University of Wollongong, Science and Technical Analysis Research Programme, Working Paper No. 9, December 1991. Online at: www.uow.edu.au/arts/sts/bmartin/dissent/documents.AIDS/Pascal91.html.
Pendell, Dale. *Living With Barbarians*. Sebastapol, Calif.: Wild Ginger Press, 1999.
———. *Pharmako/poeia*. San Francisco: Mercury House, 1995.
Perry, David. "A moveable feast: The evolution of resource sharing in plant-fungus communities," *TREE* 13, no. 11, November 1998.
Phillips, Lord, June Bridgeman, and Malcom Ferguson-Smith. *The BSE Inquiry: Vols I–XVI*. London: The Stationary Office, 2000.
Powell, W.J. "Molecular mechanisms of antimicrobial resistance," Technical Report 14, February 2000, Online at www.plant.uoguelph.ca/ safefood/ab-res-ppr-wjp.htm.
Putnam, Alan, and Chung-shih Tang. "Allelopathy: The state of the science," in

The Science of Allelopathy, ed. Alan Putnam and Chung-shih Tang. New York: John Wiley and Sons, 1986.
———, and Chung-shih Tang, eds. *The Science of Allelopathy*. New York: John Wiley and Sons, 1986.
———, and Leslie Weston. "Adverse impact of allelopathy in agricultural systems," in *The Science of Allelopathy,* ed. Alan Putnam and Chung-shih Tang. New York: John Wiley and Sons, 1986.
Rajak, R.C., and M.K. Rai. *Herbal Medicines, Bio-Diversity and Conservation Strategies.* Dehra Dun, India: International Book Distributors, 1996.
Raup, David. "The palaeobiological revolution," *TREE* 13, no. 3, March, 1998.
Reinhardt, C., et al. "Implications of allelopathic weeds for sustainability in agronomy, agroforestry, and forestry," in *Biodiversity and Allelopathy,* ed. Chang-Hung Chou, George Waller, and Charlie Reinhardt. Taipei, Taiwan: Academia Sinica, 1999.
Rodriguez, Eloy, and Richard Wrangham. "Zoopharmacognosy: The use of medicinal plants by animals," in *Phytochemical Potential of Tropical Plants,* ed. K.R. Downum, et al. New York: Plenum Press, 1993.
Romeo, John. "Functional multiplicity among nonprotein amino acids in Mimosoid legumes: A case against redundancy," *Ecoscience* 5, no. 3, 1998.
Roszak, Theodore. "Eco-Psychology," *Earth Ethics* 5, no. 2, winter 1994.
Salgado, N.R., et al. "Zoopharmacognosy of breeding neotropical birds: Chemical ecology of nest materials and substrates," Zoopharmacognosy abstracts online, www.bhort.bh.cornell.edu.
Schmidt, Michael, et al., *Beyond Antibiotics*. Berkeley, Calif.: North Atlantic Press, 1994.
Scholes, R.J., et al. "Plant functional types in African savannas and grasslands," in *Plant Functional Types: Their Relevance to Ecosystem Properties and Global Change,* ed. by T.M. Smith, H.H Shugart, and F.I. Woodward. Cambridge: Cambridge University Press, 1997.
Sheldrake, Rupert. *Morphic Resonance and the Presence of the Past*. Rochester, Vt.: Park Street Press, 1995.
———. *The Rebirth of Nature: The Greening of Science and God.* New York: Bantam, 1991.
Shugart, H.H. "Plant and ecosystem functional types," in *Plant Functional Types: Their Relevance to Ecosystem Properties and Global Change,* ed. T.M. Smith, H.H. Shugart, and F.I. Woodward, London: Cambridge University Press, 1997.
Silvertown, Jonathan. "Plant phenotypic plasticity and non-cognitive behaviour," *TREE* 13, no. 7, July 1998.
Sitwell, Edith. *Collected Poems*. London: Macmillan, 1965.
Smith, T.M., H.H. Shugart, and F.I. Woodward, eds. *Plant Functional Types: Their Relevance to Ecosystem Properties and Global Change.* London: Cambridge University Press, 1997.
Spencer, Kevin. "Chemical mediation of coevolution in the *Passiflora-Heliconium* interaction," in *Chemical Mediation of Coevolution,* ed. Kevin Spencer. New York: Academic Press, 1988.
———. "Chemistry and coevolution," in *Chemical Mediation of Coevolution,* ed. Kevin Spencer. New York: Academic Press, 1988.

———. "The chemistry of coevolution," in *Chemical Mediation of Coevolution,* ed. Kevin Spencer. New York: Academic Press, 1988.

———, ed. *Chemical Mediation of Coevolution*. New York: Academic Press, 1988.

Stevens, Kenneth. "Polyacetylenes as allelochemicals," in *The Science of Allelopathy*, ed. Alan Putnam and Chung-shih Tang, New York: John Wiley and Sons, 1986.

Stone, Christopher. *Should Trees Have Standing?: Toward Legal Rights for Natural Objects*. Los Altos, Calif.: William Kaufmann, 1974.

Stowe, Mark. "Chemical mimicry," in *Chemical Mediation of Coevolution,* ed. Kevin Spencer. New York: Academic Press, 1988.

———, et al. "The chemistry of eavesdropping, alarm, and deceit," in *Chemical Ecology: The Chemistry of Biotic Interaction,* ed. Thomas Eisner and Jerrold Meinwald. Washington D.C.: National Academy of Sciences, 1995.

Strauss, Sharon, and Anurag Agarwal, "The ecology and evolution of plant tolerance to herbivory," *TREE* 14, no. 5, May 1999.

Sumner, Judith. *The Natural History of Medicinal Plants*. Portland, Oreg.: Timber Press, 2000.

Tamogami, Shigeru, and Osamu Kodamal. "Jasmonic acid elicits momilactone A production: Physiology and chemistry of jasmonic acid in rice phytoalexin production," in *Biodiversity and Allelopathy,* ed. Chang-Hung Chou, George Waller, and Charlie Reinhardt. Taipai, Taiwan: Academia Sinica, 1999.

Tang, Chung-Shih. "Continuous trapping techniques for the study of allelochemicals from higher plants," in *The Science of Allelopathy,* ed. Alan Putnam and Chung-shih Tang. New York: John Wiley and Sons, 1986.

Taplin, Kim. *Tongues in Trees*. Devon, England: Green Books, 1989.

Tian, G., et al. "Soil fauna-mediated decomposition of plant residues under constrained environmental and residue quality conditions," in *Driven By Nature: Plant Litter Quality and Decomposition,* ed. G. Cadisch and K.E. Giller. Wallingford, England: CAB International, 1997.

Tickner, Joel, et al. *The Use of Di-2Ethylhexyl Phthalate in PVC Medical Devices: Exposure, Toxicity, and Alternatives*. Lowell, Mass.: Lowell Center for Sustainable Production, 1999.

Tompkins, Peter, and Christopher Bird. *The Secret Life of Plants*. New York: Harper and Row, 1973.

Turlings, Ted, and Betty Benrey. "Effect of plant metabolites on the behavior and development of parasitic wasps," *Ecoscience* 5, no. 3, 1998.

Tuxill, John. *Nature's Cornucopia: Our Stake in Plant Diversity*. World Watch Paper 148. Washington, D.C.: Worldwatch Institute, September 1999.

van Elsas, Jan Dirk, and Kornelia Smalla. "Antibiotic (kanamycin and streptomycin) resistance traits in the environment," *Proceedings of the Workshop on Key Biosafety Aspects of Genetically Modified Organisms*. Braunschweig, Germany, April 10–11, 1995, www.bba.de/gentech/workshoop.htm.

Wachendorf, C., et al. "Relationships between litter fauna and chemical changes of litter during decomposition under different moisture conditions," in *Driven By Nature: Plant Litter Quality and Decomposition,* ed. G. Cadisch and K. E. Giller. Wallingford, England: CAB International, 1997.

Wakeford, Tom. "Undressing the evolutionary emperor," *TREE* 13, no. 5, May 1998.

Walker, Alice. *Living by the Word*. New York: Harcourt Brace, 1988.
Walker, B.H. "Functional types in non-equilibrium ecosystems," in *Plant Functional Types: Their Relevance to Ecosystem Properties and Global Change*, ed. T.M. Smith, H.H. Shugart, and F.I. Woodward. London: Cambridge University Press, 1997.
Waller, G.R., and F.A. Einhellig. "Overview of allelopathy in agriculture, forestry, and ecology," in *Biodiversity and Allelopathy,* ed. Chang-Hung Chou, George Waller, and Charlie Reinhardt. Taipei, Taiwan: Academia Sinica, 1999.
Wardle, D.A., and P. Lavelle. "Linkages between soil biota, plant litter quality, and decomposition," in *Driven By Nature: Plant Litter Quality and Decomposition,* ed. G. Cadisch and K.E. Giller. Wallingford, England: CAB International, 1997.
Watkinson, A.R. "The role of the soil community in plant population dynamics," *TREE* 13, no. 5, May 1998.
Weathers, Kathleen. "The importance of cloud and fog in the maintenance of ecosystems," *TREE* 14, no. 6, June 1999.
West, Neil, and John Skujins. *Nitrogen in Desert Ecosystems*. Stroudsburg, Pa.: Dowden, Hutchinson and Ross, 1978.
Wilkinson, David. "Ants, agriculture, and antibiotics," *TREE* 14, no. 12, December 1999.
Wilson, Edward O. "Biophilia and the Conservation Ethic," in *The Biophilia Hypothesis,* ed. Edward O. Wilson and Stephen Kellert. Washington, D.C.: Island Press, 1993.
———., and Stephen Kellert, eds. *The Biophilia Hypothesis*. Washington, D.C.: Island Press, 1993.
Windsor, Donald. "Equal rights for parasites," *Conservation Biology* 9, no. 1 February 1995.

Resources

Stephen Harrod Buhner. Tax-deductible donations to support research and education about the living intelligence of plants and Earth (or to contact Stephen for classes or lectures): The Foundation for Gaian Studies, 612 South 11th Street, Coeur d'Alene, Idaho 83814, 208-676-9490, trishuwa@ aol.com.

Rosemary Gladstar. Tax-deductible donations to support the restoration and protection of North American medicinal plants made out to United Plant Savers, P.O. Box 98, E. Barre, VT 05649. 802-479-9825. To contact Rosemary Gladstar for classes and training in the uses of plant medicines: Sage Mountain, P.O.Box 420, E. Barre, VT 05649, 802-479-9825, www.sagemountain.com.

Carol McGrath. To contact Carol about classes and training in plant relationship and herbal medicine: 836 Swan Street, Victoria, B.C. Canada, V8X 2Z3, 250-475-3730, camcgrath@pacificcoast.net.

John Seed. Tax-deductible donations to support the protection of ancient forests throughout the world made out to: Rainforest Information Centre, Box 368, Lismore, New South Wales 2480, Australia. To find out about John Seed programs or to attend classes: johnseed@ozemail.com.au, http://forests.org/ric/; 61-2-66213294.

Sparrow (Jonathan Sparrow Miller). Tax-deductible donations to support the protection of traditional medicinal knowledge and territory in Ecuador made out to: Grupo Osanimi, P.O. Box 1004, El Cerrito, CA 94530. To contact Sparrow for trips to Ecuador: Sentient Experientials, Dahlia Miller, 510-235-4313, sentient@experientials.org, www.experientials.org.

Index

E

F

G

H

Q

R

S

T

U

V

W

X

Y

Z

Sustainable living has many facets. Chelsea Green's celebration of the sustainable arts has led us to publish trend-setting books about organic gardening, solar electricity and renewable energy, innovative building techniques, regenerative forestry, local and bioregional democracy, and whole foods. The company's published works, while intensely practical, are also entertaining and inspirational, demonstrating that an ecological approach to life is consistent with producing beautiful, eloquent, and useful books, videos, and audio cassettes.

For more information about Chelsea Green, or to request a free catalog, call toll-free (800) 639-4099, or write to us at P.O. Box 428, White River Junction, Vermont 05001. Visit our Web site at www.chelseagreen.com.

Chelsea Green's titles include:

The Natural House
The Straw Bale House
The New Independent Home
The Hand-Sculpted House
Serious Straw Bale
The Beauty of Straw Bale Homes
Building with Earth
The Resourceful Renovator
Independent Builder
The Rammed Earth House
The Passive Solar House
Wind Energy Basics
Wind Power for Home & Business
The Solar Living Sourcebook
A Shelter Sketchbook
Mortgage-Free!
Stone Circles

Gaia's Garden
The Neighborhood Forager
The Apple Grower
The Flower Farmer
Breed Your Own Vegetable Varieties
Keeping Food Fresh
The Soul of Soil
The New Organic Grower
Four-Season Harvest
Solar Gardening
Straight-Ahead Organic
The Contrary Farmer
The Co-op Cookbook
Whole Foods Companion
The Bread Builder
The Village Herbalist
The Pesto Manifesto
Secrets of Salsa

A Cafecito Story
Believing Cassandra
Gaviotas: A Village to Reinvent the World
Who Owns the Sun?
Global Spin: The Corporate Assault on Environmentalism
Hemp Horizons
This Organic Life
Beyond the Limits
The Man Who Planted Trees
The Northern Forest
The New Settler Interviews
Loving and Leaving the Good Life
Scott Nearing: The Making of a Homesteader
Wise Words for the Good Life